Finite Element Methods in Civil and Mechanical Engineering

Finite Element Methods in Civil and Mechanical Engineering

A Mathematical Introduction

Dr Arzhang Angoshtari
Ali Gerami Matin

CRC Press
Taylor & Francis Group
Boca Raton London New York

CRC Press is an imprint of the
Taylor & Francis Group, an **informa** business

First edition published 2021
by CRC Press
6000 Broken Sound Parkway NW, Suite 300, Boca Raton, FL 33487-2742

and by CRC Press
2 Park Square, Milton Park, Abingdon, Oxon, OX14 4RN

© 2021 Taylor & Francis Group, LLC

CRC Press is an imprint of Taylor & Francis Group, LLC

Library of Congress Cataloging-in-Publication Data

Names: Angoshtari, Arzhang, author. | Matin, Ali Gerami, author.
Title: Finite element methods in civil and mechanical engineering : a
    mathematical introduction / Arzhang Angoshtari, Ali Gerami Matin.
Description: Boca Raton : CRC Press, 2020. | Includes bibliographical
    references and index.
Identifiers: LCCN 2020030081 (print) | LCCN 2020030082 (ebook) | ISBN
    9781138335172 (paperback) | ISBN 9781138335165 (hardback) | ISBN
    9780429442506 (ebook)
Subjects: LCSH: Finite element method. | Civil engineering--Mathematics. |
    Mechanical engineering--Mathematics.
Classification: LCC TA347.F5 A54 2020 (print) | LCC TA347.F5 (ebook) |
    DDC 518/.25--dc23
LC record available at https://lccn.loc.gov/2020030081
LC ebook record available at https://lccn.loc.gov/2020030082

ISBN: 9781138335165 (hbk)
ISBN: 9781138335172 (pbk)
ISBN: 9780429442506 (ebk)

Typeset in Computer Modern font
by KnowledgeWorks Global Ltd.

by Arzhang Angoshtari

Visit the eResource: www.routledge.com/9781138335165

Dedication

To Our Parents

Contents

Preface xi

Chapter 1   Overview 1

Chapter 2   Mathematical Preliminaries 5

     2.1 Real Numbers 5
     2.2 Functions 5
     2.3 Linear Spaces, Linear Mappings, and Bilinear Forms 6
     2.4 Linear Independence, Hamel Bases, and Dimension 9
     2.5 The Matrix Representation of Linear Mappings and
           Bilinear Forms 10
     2.6 Normed Linear Spaces 12
     2.7 Functionals and Dual Spaces 13
     2.8 Green's Formulas 14
     Exercises 15
     Comments and References 17

Chapter 3   Finite Element Interpolation 19

     3.1 1D Finite Element Interpolation 19
           3.1.1 The Global Level 19
           3.1.2 The Local Level 24
     3.2 Finite Elements 26
           3.2.1 Simplicial Lagrange Finite Elements of Type (k) 26
           3.2.2 Simplicial Hermite Finite Elements of Type (3) 30
           3.2.3 The Raviart-Thomas Finite Element 31
           3.2.4 The Nedelec Finite Element 33
     3.3 Meshes 34
     3.4 Finite Element Spaces and Interpolations 37
           3.4.1 H^1-Conformal Finite Element Spaces 38
                3.4.1.1 Lagrange Elements 39
                3.4.1.2 Hermite Elements 43
           3.4.2 $H(\mathrm{div})$-Conformal Finite Element Spaces 43
           3.4.3 $H(\mathrm{curl})$-Conformal Finite Element Spaces 45
           3.4.4 Affine Families of Finite Elements 45
     3.5 Convergence of Interpolations 46
     Exercises 50
     Computer Exercises 51

Comments and References...52

Chapter 4   Conforming Finite Element Methods for PDEs 55

4.1   Second-Order Elliptic PDEs...55
4.2   Weak Formulations of Elliptic PDEs...................................56
    4.2.1   Dirichlet Boundary Condition................................57
    4.2.2   Neumann Boundary Condition 58
    4.2.3   Robin Boundary Condition 59
4.3   Well-posedness of Weak Formulations...............................60
4.4   Variational Structure 61
4.5   The Galerkin Method and Finite Element Methods 62
    4.5.1   The Stiffness Matrix 63
    4.5.2   Well-posedness of Coercive Discrete Problems 64
    4.5.3   Convergence of Finite Element Solutions...............64
4.6   Implementation: The Poisson Equation..............................66
    4.6.1   Dirichlet Boundary Condition................................66
    4.6.2   Mixed Dirichlet-Neumann Boundary Condition 70
    4.6.3   Robin Boundary Condition 74
4.7   Time-Dependent Problems: Parabolic Problems.................76
    4.7.1   Finite Element Approximations using the Method
        of Lines 78
    4.7.2   Temporal Discretization...79
    4.7.3   Implementation: A Diffusion Problem 79
4.8   Mixed Finite Element Methods 82
    4.8.1   Mixed Formulations..83
    4.8.2   Mixed Methods and inf-sup Conditions 84
    4.8.3   Implementation 85
Exercises 90
Computer Exercises 92
Comments and References...93

Chapter 5   Applications..97

5.1   Elastic Bars 97
5.2   Euler-Bernoulli Beams 100
5.3   Elastic Membranes..104
5.4   The Wave Equation 105
5.5   Heat Transfer in a Turbine Blade......................................107
5.6   Seepage in Embankment..112
5.7   Soil Consolidation 117
5.8   The Stokes Equation for Incompressible Fluids................122
5.9   Linearized Elasticity 126
5.10 Linearized Elastodynamics: The Hamburg Wheel-Track
      Test..129
5.11 Nonlinear Elasticity 134

Exercises 143
Computer Exercises 145

Appendix A   FEniCS Installation 147

Appendix B   Introduction to Python.. 149
B.1   Running Python Programs............................... 149
B.2   Lists.. 150
B.3   Branching and Loops....................................... 151
B.4   Functions... 152
B.5   Classes and Objects 152
B.6   Reading and Writing Files 154
B.7   Numerical Python Arrays 155
B.8   Plotting with Matplotlib.................................. 157

References...**159**

Index..**161**

Preface

In the past century, the finite element method has grown significantly and nowadays it is considered as a standard tool for dealing with many engineering and scientific problems. Numerous computer programs are available today for the implementation of the finite element method, which relieve one from implementation of the main aspects of the finite element method from scratch.

Significant contributions to this field have been made by engineers and mathematicians. Methods developed by engineers are sometimes suitable for specific applications or circumstances and cannot be generalized easily. They may lead to confusions when applied to new applications, some of which can be addressed by using rich results already available in the mathematical literature. However, mathematical results are usually not easily accessible to many engineers due to their rigorous nature.

The main goal of this book is to provide a modern and concise introduction to the theory and the implementation of the finite element method for engineers and physicists. The emphasis is on the fundamental aspects of the finite element method for solving linear partial differential equations. We use a mathematical language to introduce main concepts. Although this approach may look cumbersome at the beginning, it is very useful and powerful not only because it provides a suitable framework to develop, implement, and study finite element methods but also because it can provide sufficient background and self-confidence to the readers to refer to the mathematical literature. We use a Python based package called FEniCS to implement finite element methods. FEniCS syntax is close to the mathematical language and it yields concise, readable, and efficient programs.

This book is based on the graduate-level course that the first author taught at the George Washington University in 2016–2019. It is suitable for a one-semester graduate course in engineering and physics. It may be used in a course for advanced undergraduate students in mathematics as well. The reader is assumed to be familiar with elementary calculus and basic programming skills. The book is organized as follows:

Chapter 1 contains an overview of the topics that will be discussed in the book.

In Chapter 2, mathematical preliminaries are briefly introduced. We only discuss main ideas and results. Although this shallow preliminary chapter is far from being complete, based on our own experience with graduate engineering students, we believe it enables the readers who do not have solid background in mathematics to understand the main ideas and refer to suitable references when necessary.

Finite elements, finite element spaces, and finite element interpolations are discussed in Chapter 3. In addition to Lagrange and Hermite finite elements, we also mention the so-called edge and face elements for the curl and div operators. The convergence of finite element interpolations is briefly discussed as well.

In Chapter 4, we study the application of finite elements for solving specific classes of time-independent and time-dependent partial differential equations that arise in many applications. We mention basic results on the well-posedness and the convergence of finite element methods. A brief discussion of mixed finite element methods and inf-sup conditions are also given in this chapter. Moreover, FEniCS implementations of finite element methods for various boundary conditions are provided.

In Chapter 5, a selection of applications in civil and mechanical engineering is studied and Python programs for their implementation are provided.

Numerous examples, exercises, and computer exercises are given in each chapter. Additional comments and suggested references for further reading are mentioned at the end of the chapters. Several Python programs for the implementation of examples are provided throughout the book. Previous familiarity with Python, although very helpful, is not necessary. Two short appendices on FEniCS installation and Python are provided at the end of the book. The latter is a brief introduction intended for the readers who are not familiar with Python. For the readers convenience, Jupyter notebooks containing all Python programs of each chapter are available online on the companion website of the book.

We would like to thank Marmar Mehrabadi for her valuable feedback. We are also grateful to all people in CRC Press who helped us during the preparation of this book. A special thanks goes to our editor Tony Moore for encouraging us to write this book.

Arzhang Angoshtari, Ali Gerami Matin
Washington, DC
May 2020

1 Overview

In this introductory chapter, we briefly discuss the basic aspects of the finite element method for solving differential equations. The goal is to give an overview of the book and to motivate the topics that will be discussed in the following chapters. Only main ideas are discussed and the readers are expected to focus only on the big picture and not on details. Technical details of this chapter are the subject of the following ones. This chapter is not a prerequisite for the rest of the book and the readers may skip this chapter if they wish.

Roughly speaking, one may consider the following four stages for the approximation of responses of a system using the finite element method: (i) Identifying the governing equations, which are assumed to be differential equations in this book; (ii) Deriving weak forms of the governing equations; (iii) Discretizing weak forms by using the Galerkin method to obtain a discrete problem; and (iv) Employing piecewise polynomial finite element spaces to solve the discrete problem.

To be more specific, let us consider the approximation of the equilibrium position of an elastic membrane Ω shown in Figure 1.1, which is fixed at its boundary $\partial\Omega$ and is under a vertical load f.

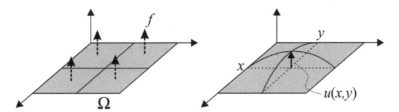

Figure 1.1 An elastic membrane under a vertical load: (Left) Undeformed configuration Ω and the load f (dashed arrows); (Right) The vertical displacement $u(x,y)$ at the point (x,y).

Governing Equations

The laws of physics imply that the vertical displacement u of the membrane Ω satisfies the boundary value problem

$$\begin{cases} -\Delta u = f, & \text{in } \Omega, \\ u = 0, & \text{on } \partial\Omega, \end{cases} \tag{1.1}$$

where Δ is *the Laplacian*, i.e. $\Delta u = \partial_{xx}u + \partial_{yy}u$. This *governing equation* is also called *Poisson's equation*.

Weak Forms

The equation (1.1) is called the *strong form* of the governing equation. To solve the strong form, one should find a function which is at least twice differentiable and satisfies (1.1). Usually, it is hard to directly find such a function. Alternatively, it may be easier to find a suitable solution candidate first and then showing that the solution candidate is a real solution by establishing its differentiability.

We can obtain a solution candidate for the strong form (1.1) by using one of its *weak forms*. For example, let $w(x,y)$ be an arbitrary function that vanishes on the boundary. Multiplying the strong form by w and then taking the integral over Ω, one concludes that

$$\int_{\Omega} \left\{ \partial_x u\, \partial_x w + \partial_y u\, \partial_y w \right\} dx\,dy = \int_{\Omega} f w\, dx\,dy, \quad \text{for all } w. \tag{1.2}$$

A solution u of (1.1) is also a solution of (1.2), however, the converse may not be true in general. Notice that unlike a solution of (1.1) that should be twice differentiable, a solution of (1.2) may be merely once differentiable since only the first derivatives of u are present in (1.2). The equation (1.2) is called a *weak form* of (1.1) due to this weaker differentiability requirement.

A solution of (1.2) can be considered as a solution candidate for (1.1). For computational purposes, weak forms are usually more suitable than strong forms.

The Galerkin Method

Since finding an analytical solution for the weak form is not easy in general, we can try to approximate its solution. *The Galerkin method* provides a systematic framework for this purpose. More specifically, given a set of functions $\{ \psi_1, \ldots, \psi_N \}$ called *basis functions* or *global shape functions*, one assumes that a solution of the weak form can be approximated by using a function u_h of the form

$$u_h(x,y) = \sum_{i=1}^{N} U_i \psi_i(x,y),$$

where U_i, $i = 1, \ldots, N$, are unknown constants. To determine the constants U_i, we assume u_h is a solution of the following problem, which is a special case of (1.2) with $w = \psi_1, \ldots, \psi_N$:

$$\int_{\Omega} \left\{ \partial_x u_h\, \partial_x \psi_i + \partial_y u_h\, \partial_y \psi_i \right\} dx\,dy = \int_{\Omega} f \psi_i\, dx\,dy, \quad i = 1, \ldots, N. \tag{1.3}$$

The above framework for approximating a solution of the weak form (1.2) is called the Galerkin method. The problem (1.3) is called a *discrete* problem or a *discretization* of (1.2) and u_h is called a *discrete* solution in the sense that the approximate solution u_h is determined by a finite number of unknowns U_i, $i = 1, \ldots, N$.

The discrete problem (1.3) is very suitable for numerical analysis. In fact, it is equivalent to the linear system of equations

$$\mathbb{A}_{N \times N} \cdot \mathbb{U}_{N \times 1} = \mathbb{F}_{N \times 1}, \tag{1.4}$$

where

$$\mathbb{A} = \begin{bmatrix} A(\psi_1,\psi_1) & A(\psi_2,\psi_1) & \cdots & A(\psi_N,\psi_1) \\ A(\psi_1,\psi_2) & A(\psi_2,\psi_2) & \cdots & A(\psi_N,\psi_2) \\ \vdots & \vdots & & \vdots \\ A(\psi_1,\psi_N) & A(\psi_2,\psi_N) & \cdots & A(\psi_N,\psi_N) \end{bmatrix}, \quad \mathbb{U} = \begin{bmatrix} U_1 \\ U_2 \\ \vdots \\ U_N \end{bmatrix}, \quad \mathbb{F} = \begin{bmatrix} F(\psi_1) \\ F(\psi_2) \\ \vdots \\ F(\psi_N) \end{bmatrix},$$

with

$$A(\psi_j,\psi_i) = \int_\Omega \left\{ \partial_x \psi_j \, \partial_x \psi_j + \partial_y \psi_j \, \partial_y \psi_i \right\} dx\, dy,$$
$$F(\psi_i) = \int_\Omega f\psi_i \, dx\, dy. \tag{1.5}$$

The matrix $\mathbb{A}$ is usually called *the stiffness matrix* and the vector $\mathbb{F}$ is called *the load vector*. In summary, the Galerkin method for approximating a solution of (1.2) results in solving the linear system (1.4) for obtaining the unknown coefficients U_i.

The Finite Element Method

We saw that once a set of basis functions $\{\psi_1,\ldots,\psi_N\}$ is chosen, one can apply the Galerkin method to (1.2) to approximate a solution of (1.1). But how can we determine a suitable choice of basis functions? There are many different approaches for obtaining basis functions which yield different approximation methods such as finite element methods, finite-difference methods, spectral methods, etc. We will study finite element methods in this book. The main aspects of the finite element method for obtaining basis functions can be summarized as follows.

(i) A *mesh* (also called a *triangulation*) T_h is established over Ω. Each mesh consists of several cells or elements that can be, in principle, of arbitrary shape, see Figure 1.2.

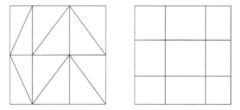

Figure 1.2 A triangular and a rectangular mesh of the rectangular membrane Ω.

(ii) Given a mesh T_h, basis functions are chosen such that their restriction to any element of T_h are polynomials or "close to" polynomials. The convergence of finite element methods is closely related to this aspect. Moreover, this leads to simple and efficient computations of stiffness matrices and load vectors.

(iii) Basis functions are selected to be *locally supported*, i.e., they are nonzero only on a few elements of T_h. For example, given the triangular mesh of Figure 1.2, one may select the basis functions ψ_1 and ψ_2 such that they are zero everywhere except on a few elements shown in Figure 1.3. Then, the definition (1.5) of the entries of the stiffness matrix $\mathbb{A}$ implies that $A(\psi_1, \psi_2) = A(\psi_2, \psi_1) = 0$. Due to this aspect of the finite element method, the stiffness matrix $\mathbb{A}$ is *sparse*, i.e., most of the entries of $\mathbb{A}$ are zero. The sparsity of $\mathbb{A}$ allows us to efficiently compute the solution of the linear system (1.4).

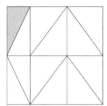
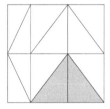

Figure 1.3 The shaded elements depict regions where the shape functions ψ_1 (left) and ψ_2 (right) are nonzero.

Implementation. Based on the Galerkin method, the finite element method provides a computationally efficient framework for approximating solutions of partial differential equations by using locally supported, piecewise polynomial basis functions. In this book, we will employ FEniCS, a Python package, to implement finite element methods. An interesting feature of FEniCS is that its syntax is close to the mathematical notion. Figure 1.4 shows a finite element approximation of a deformation of the elastic membrane of Figure 1.1 under a constant vertical load.

In the remainder of this book, we will study the ideas and notions of this chapter in more details.

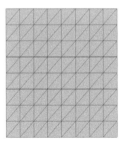
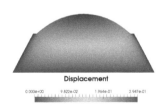

Figure 1.4 A finite element approximation of a deformation of the elastic membrane Ω of Figure 1.1 under a constant vertical load: (Left) The underlying mesh; (Right) the final configuration of Ω.

2 Mathematical Preliminaries

Normed linear spaces and linear operators play a central role in the finite element method. In this chapter, we study these notions and also fix our notations. We will only mention main ideas in a level sufficient for understanding the discussions of the following chapters. For more examples and discussions on technical details, we refer the reader to the references mentioned at the end of this chapter.

2.1 REAL NUMBERS

The set of real numbers is denoted by $\mathbb{R}$. For any real numbers a, b with $a < b$, (a, b) represents the open interval $\{x \in \mathbb{R} : a < x < b\}$ and $[a, b]$ represents the closed interval $\{x \in \mathbb{R} : a \leq x \leq b\}$. By abusing the notation, we also use (a, b) to denote a point in the plane $\mathbb{R}^2$ but the meaning is always clear from the context.

The empty set is denoted by $\varnothing$. Suppose $A \neq \varnothing$ is a non-empty subset of $\mathbb{R}$. We say that A is *bounded from above* if there is a number u such that $a \leq u$, for all $a \in A$, and we call u an *upper bound* of A. We say A is *bounded from below* if there is a number l such that $a \geq l$, for all $a \in A$, and we call l a *lower bound* of A. The set A is called *bounded* if it is bounded from below and above. The *maximum* of $A \subset \mathbb{R}$, denoted by $\max A$, is a member of A which is also an upper bound of A. Similarly, the *minimum* of $A \subset \mathbb{R}$, denoted by $\min A$, is a member of A which is also a lower bound of A. For example, $\max[0, 1] = 1$ and $\min[0, 1] = 0$.

Notice that the maximum and the minimum of a bounded set may not exist, for example, consider the open interval $(0, 1)$. A basic fact about $\mathbb{R}$ allows one to extend the notions of maximum and minimum to all bounded subsets in the following sense: If $B \subset \mathbb{R}$ is non-empty and is bounded from above, then there is the least upper bound of B, which is called the *supremum* of B and is denoted by $\sup B$. Similarly, if a non-empty set B is bounded from below, then there is the greatest lower bound of B, which is called the *infimum* of B and is denoted by $\inf B$. For example, $\inf(0, 1) = 0$, and $\sup(0, 1) = 1$. If $\max B$ exits, then $\max B = \sup B$, and if $\min B$ exists, then $\min B = \inf B$.

2.2 FUNCTIONS

Let X and Y be two sets. Recall that a function (also called a mapping or an operator) $f : X \to Y$ assigns a unique value $f(x) \in Y$ to each $x \in X$. The function f is called a Y-valued function. The set X is called the *domain* of f and is also denoted by $\mathrm{D}(f)$. A function g is called an *extension* of f if $\mathrm{D}(f) \subset \mathrm{D}(g)$, and $f(x) = g(x)$, for all $x \in \mathrm{D}(f)$. In this case, the function f is called a *restriction* of g and we write $f = g|_{\mathrm{D}(f)}$.

Example 2.1. The function $g : \mathbb{R} \to \mathbb{R}$, $g(x) = |x|$, is an extension of $f : (0, 1) \to \mathbb{R}$, $f(x) = x$, or equivalently, we have $f = g|_{(0,1)}$.

The *range* of $f : X \to Y$ is a subset of Y defined as

$$\mathrm{R}(f) = \{y \in Y : \text{there is } x \in X \text{ with } y = f(x)\}.$$

If $\mathrm{R}(f) = Y$, f is called *onto* or *surjective*. A function $f : X \to Y$ is called *one-to-one* or *injective* if different points do not take the same values, that is, $f(x_1) = f(x_2)$, implies that $x_1 = x_2$. A mapping which is injective and surjective is called *bijective*. A bijective function $f : X \to Y$ is *invertible*, that is, f admits the inverse function $f^{-1} : Y \to X$, where for any $x \in X$ and $y \in Y$, we have $f^{-1}(f(x)) = x$, and $f(f^{-1}(y)) = y$.

Example 2.2. The function $f : (0,1) \to \mathbb{R}$, $f(x) = x$, is injective but not surjective, while $g : (0,1) \to (0,1)$, $g(x) = x$, is bijective.

Suppose $A \subset X$ and $B \subset Y$. For any arbitrary function $f : X \to Y$, the sets $f(A) \subset Y$ and $f^{-1}(B) \subset X$ are defined as $f(A) = \{y \in Y : y = f(x) \text{ for some } x \in A\}$, and $f^{-1}(B) = \{x \in X : f(x) \in B\}$. Notice that for defining the set $f^{-1}(B)$, f does not need to be invertible.

Example 2.3. Let $f : \mathbb{R} \to \mathbb{R}$, $f(x) = x^2$. Then, $f([0,2]) = [0,4]$, and $f^{-1}([-2,1]) = [-1,1]$.

Roughly speaking, we say Ω is an *open* subset of the Euclidean space $\mathbb{R}^n$ if Ω is n-dimensional and it does not contain its boundary. The *closure* $\overline{\Omega}$ of the open subset Ω is the union of Ω and the set of its *boundary points* $\partial\Omega$. For example, let *the unit ball* $B_1(0)$ be the set of all points of $\mathbb{R}^3$ such that their distance from the origin 0 is smaller than 1. Then, $B_1(0)$ is an open subset of $\mathbb{R}^3$ and $\overline{B_1(0)}$ is the union of the unit ball and its boundary, which is *the unit sphere* with radius 1 centered at the origin.

Let $\Omega \subset \mathbb{R}^n$ be open. The set of all functions $u : \Omega \to \mathbb{R}$ with continuous derivatives up to order m is denoted by $C^m(\Omega)$. For example, suppose $\Omega \subset \mathbb{R}^2$. Then, $C^0(\Omega)$ is the set of continuous functions on Ω and $C^2(\Omega)$ is the set of functions u such that $u, \partial_x u, \partial_y u, \partial_{xx} u, \partial_{xy} u, \partial_{yy} u$ are continuous on Ω, where $\partial_x u$ is the partial derivative with respect to x and so on. The set $C^m(\overline{\Omega})$ is the set of functions $u \in C^m(\Omega)$ such that u and all of its partial derivatives up to order m can be continuously extended to $\overline{\Omega}$. Notice that $C^m(\overline{\Omega}) \subset C^m(\Omega)$ but $C^m(\Omega) \not\subset C^m(\overline{\Omega})$.

Example 2.4. Let $\Omega = (0,1)$, with $\overline{\Omega} = [0,1]$, and consider $f(x) = 1/x$, and $g(x) = x^2$. Then, $f \in C^1(\Omega)$ but $f \notin C^1(\overline{\Omega})$, while g belongs to both $C^1(\Omega)$ and $C^1(\overline{\Omega})$.

2.3 LINEAR SPACES, LINEAR MAPPINGS, AND BILINEAR FORMS

A *linear space*, also called a *vector space*, is a set X together with an *addition* and a *scalar multiplication*. Let $x, y, z \in X$ and $a, b \in \mathbb{R}$. The addition and the scalar multiplication, respectively denoted by $x + y$ and ax, are assumed to satisfy the following conditions:

- The addition of any two members of X is another member of X such that $x + y = y + x$, and $x + (y + z) = (x + y) + z$;

- There is a unique *zero* member $0 \in X$ with $0 + x = x$. Also for any x, there is a unique member $-x \in X$ such that $x + (-x) = 0$;
- The scalar multiplication satisfies $1x = x$, $0x = 0$, and $a(bx) = (ab)x$;
- The addition and the scalar multiplication satisfy $a(x + y) = ax + ay$, and $(a + b)x = ax + bx$.

For brevity, we only mention a linear space X when its addition and scalar multiplication are known. A simple example of a linear space is the Euclidean space $\mathbb{R}^n$ with its standard addition and scalar multiplication. Less trivial linear spaces which are frequently encountered in finite element methods are discussed in the following examples.

Example 2.5. Given an open subset $\Omega \subset \mathbb{R}^n$, the space $C^m(\Omega)$ is a linear space: For any $f, g \in C^m(\Omega)$, and $c \in \mathbb{R}$, the addition and the scalar multiplication are defined pointwise as $(f + g)(x) = f(x) + g(x)$, and $(cf)(x) = cf(x)$, for all $x \in \Omega$. Similarly, one can show that $C^m(\overline{\Omega})$ is a linear space.

Example 2.6. Let $\mathbb{P}_k(\mathbb{R}^n)$ denote the space of all polynomials of degree $\leq k$ in $\mathbb{R}^n$. For example, $\mathbb{P}_2(\mathbb{R})$ is the space of polynomials of the form $c_0 + c_1 x + c_2 x^2$, where c_i, $i = 0, 1, 2$, are arbitrary constants, and $\mathbb{P}_2(\mathbb{R}^2)$ is the space of polynomials of the form $c_{00} + c_{10}x + c_{01}y + c_{20}x^2 + c_{11}xy + c_{02}y^2$, where c_{ij}, $i, j = 0, 1, 2$, are arbitrary constants. By using the pointwise operations similar to those of the previous example, one can show that $\mathbb{P}_k(\mathbb{R}^n)$ is a linear space. For any $A \subset \mathbb{R}^n$, the set of polynomials on A defined as $\mathbb{P}_k(A) = \{p|_A : p \in \mathbb{P}_k(\mathbb{R}^n)\}$, is also a linear space.

Example 2.7 (Lebesgue spaces). For any open subset $\Omega \subset \mathbb{R}^n$, the L^2 *Lebesgue space* is defined as $L^2(\Omega) = \{f : \Omega \to \mathbb{R} : \int_\Omega |f|^2 \text{ exists}\}$, where by $\int_\Omega |f|^2$ we mean $\int_\Omega |f(x_1, \ldots, x_n)|^2 dx_1 \cdots dx_n$. For brevity, we follow this notation for integrals from now on. For example, let $\Omega = (0, 1)$, and $f(x) = x^\alpha$. Then, $f \in L^2(\Omega)$ if $\alpha > -\frac{1}{2}$. By using the pointwise operations, one can show that $L^2(\Omega)$ is a linear space. A vector field $v : \Omega \to \mathbb{R}^n$ is said to be of L^2-class if its components belong to $L^2(\Omega)$. The space of L^2 vector fields is denoted by $[L^2(\Omega)]^n$ and is a linear space.

Example 2.8 (Sobolev spaces). The *Sobolev space* $H^1(\Omega)$ is the space of functions in $L^2(\Omega)$ which also have L^2 first-order derivatives, that is, $H^1(\Omega) = \{f \in L^2(\Omega) : \partial_{x_i} f \in L^2(\Omega), i = 1, \ldots, n\}$. For example, assuming $\beta > \frac{1}{2}$, $\Omega = (0, 1)$, and $f(x) = x^\beta$, then $f \in H^1(\Omega)$. The Sobolev space of vector fields $\Omega \to \mathbb{R}^n$ with components belonging to $H^1(\Omega)$ is denoted by $[H^1(\Omega)]^n$. One can also define the Sobolev space $H^m(\Omega)$ for any positive integer m. For example, $H^2(\Omega)$ is the space of functions in $L^2(\Omega)$ with first-order and second-order derivatives also belonging to $L^2(\Omega)$. These Sobolev spaces become linear spaces when equipped with the pointwise operations.

Example 2.9 (Partly Sobolev classes). Recall that the divergence of a vector field $v : \Omega \to \mathbb{R}^n$ with components $v = (v_1, \ldots, v_n)$ is defined as $\operatorname{div} v = \sum_{i=1}^n \partial_{x_i} v_i$. The *partly Sobolev class* associated to the divergence operator is defined as $H(\operatorname{div}; \Omega) = \{v \in [L^2(\Omega)]^n : \operatorname{div} v \in L^2(\Omega)\}$. This space becomes a linear space with pointwise operations. One can also define the partly Sobolev space $H(\mathrm{curl};\Omega)$: Recall that in $\mathbb{R}^3$, the curl operator is given by

$$\mathrm{curl}\,\boldsymbol{v} = (\partial_{x_2}v_3 - \partial_{x_3}v_2, \partial_{x_3}v_1 - \partial_{x_1}v_3, \partial_{x_1}v_2 - \partial_{x_2}v_1).$$

Then, $H(\mathrm{curl};\Omega) = \{\boldsymbol{v} \in [L^2(\Omega)]^3 : \mathrm{curl}\,\boldsymbol{v} \in [L^2(\Omega)]^3\}$, is a linear space. By using the 2-dimensional curl operator $\mathrm{curl}\,\boldsymbol{v} = \partial_{x_1}v_2 - \partial_{x_2}v_1$, the partly Sobolev class $H(\mathrm{curl};\Omega)$ can be defined for 2-dimensional vector fields as well. One can show that $[H^1(\Omega)]^n \subset H(\mathrm{div};\Omega) \subset [L^2(\Omega)]^n$, and $[H^1(\Omega)]^n \subset H(\mathrm{curl};\Omega) \subset [L^2(\Omega)]^n$.

A subset S of a linear space X is called a *linear subspace* of X if with respect to the addition and the scalar multiplication of X, S is also a linear space.

Example 2.10. Any line passing through the origin is a linear subspace of $\mathbb{R}^2$.

Example 2.11. Let $\partial\Omega$ be the boundary of an open subset $\Omega \subset \mathbb{R}^n$. Then, $H_0^1(\Omega) = \{f \in H^1(\Omega) : u|_{\partial\Omega} = 0\}$, is a linear subspace of $H^1(\Omega)$. The polynomial space $\mathbb{P}_k(\Omega)$ is another subspace of $H^1(\Omega)$.

Let X and Y be linear spaces. A mapping $L : X \to Y$ is called a *linear* mapping if for any finite subset $\{x_1,\ldots,x_n\} \subset X$ and $\{c_1,\ldots,c_n\} \subset \mathbb{R}$, we have $L(\sum_{i=1}^n c_i x_i) = \sum_{i=1}^n c_i L(x_i)$.

Example 2.12. An $m \times n$ matrix M can be considered as a mapping $M : \mathbb{R}^n \to \mathbb{R}^m$, defined as $M(\mathbf{x}) = M \cdot \mathbf{x}$. This mapping is a linear mapping.

Example 2.13. For any $f \in C^0(\overline{\Omega})$, consider the mapping $L : C^0(\overline{\Omega}) \to \mathbb{R}$, given by $L(u) = \int_\Omega fu$. Since for any $u_i \in C^0(\overline{\Omega})$, and $c_i \in \mathbb{R}$, $i = 1,\ldots,n$, we have

$$L\Big(\sum_{i=1}^n c_i u_i\Big) = \int_\Omega f \cdot \Big(\sum_{i=1}^n c_i u_i\Big) = \sum_{i=1}^n c_i \int_\Omega f u_i = \sum_{i=1}^n c_i L(u_i),$$

the operator L is a linear operator.

Example 2.14. Suppose $\mathbf{a} \in \mathbb{R}^m$ and M is an $m \times n$ matrix. A mapping $T : \mathbb{R}^n \to \mathbb{R}^m$ of the form $T(\mathbf{x}) = \mathbf{a} + M \cdot \mathbf{x}$ is said to be an *affine* mapping. Affine mappings are not linear.

The null space of a linear mapping $L : X \to Y$ is defined as $\mathrm{N}(L) = L^{-1}(\{0\})$, that is,

$$\mathrm{N}(L) = \{x \in X : L(x) = 0\}.$$

The null space of L is also called the *kernel* of L and is denoted by $\ker L$. The null space of $L : X \to Y$ is a linear subspace of X and the range $\mathrm{R}(L)$ of L is a linear subspace of Y. Moreover, L is injective if and only if $\ker L = 0$.

Example 2.15. For the linear mapping $L : \mathbb{R}^3 \to \mathbb{R}^3$ given by

$$L = \begin{bmatrix} 1 & 0 & 0 \\ 0 & 0 & 0 \\ 0 & 0 & 0 \end{bmatrix},$$

the spaces $N(L)$ and $R(L)$ are respectively the y-z plane and the x-axis.

Given linear spaces X and Y, a mapping $B : X \times Y \to \mathbb{R}$ is called a *bilinear form* if for any finite subsets $\{x, x_1, \ldots, x_n\} \subset X$, $\{y, y_1, \ldots, y_m\} \subset Y$, and $\{a_1, \ldots, a_n\}, \{b_1, \ldots, b_m\} \subset \mathbb{R}$, we have

$$B\left(\sum_{i=1}^{n} a_i x_i, y\right) = \sum_{i=1}^{n} a_i B(x_i, y), \text{ and } B\left(x, \sum_{j=1}^{m} b_j y_j\right) = \sum_{j=1}^{m} b_j B(x, y_j).$$

Thus, $B(x, y)$ is a bilinear form if it is linear with respect to each of its entries. A bilinear form $B : X \times X \to \mathbb{R}$ is symmetric if $B(x_1, x_2) = B(x_2, x_1)$, for all $x_1, x_2 \in X$.

Example 2.16. An $m \times n$ matrix M can be considered as a bilinear form $B : \mathbb{R}^n \times \mathbb{R}^m \to \mathbb{R}$, given by $B(\mathbf{x}, \mathbf{y}) = \mathbf{y}^T \cdot M \cdot \mathbf{x}$, where $\mathbf{y}^T$ is the transpose of the column vector $\mathbf{y}$. Since $\mathbf{y}^T \cdot M \cdot \mathbf{x} = \mathbf{x}^T \cdot M^T \cdot \mathbf{y}$, one concludes that $M_{n \times n}$ is a symmetric bilinear form if and only if the matrix M is symmetric, that is, $M = M^T$.

Example 2.17. The mapping $B : C^0(\overline{\Omega}) \times C^0(\overline{\Omega}) \to \mathbb{R}$, $B(f, g) = \int_\Omega fg$, is a symmetric bilinear form.

2.4 LINEAR INDEPENDENCE, HAMEL BASES, AND DIMENSION

Suppose a set S, perhaps having *infinite* members, is a subset of a linear space X. A point $s \in X$ is a *linear combination* of points in S if there are *finite* points $s_1, \ldots, s_n$ in S such that $s = \sum_{i=1}^{n} c_i s_i$, $c_i \in \mathbb{R}$, $i = 1, \ldots, n$. The set of all linear combinations of points of S, also called the linear subspace *spanned* by S, is denoted by $\text{span} S$.

Example 2.18. Let $S = \{1, x, x^2\} \subset \mathbb{P}_2(\mathbb{R})$. Then $\text{span} S = \mathbb{P}_2(\mathbb{R})$.

A subset S of a linear space X is called *linearly independent* if for any *finite* collection $\{s_1, \ldots, s_n\}$ of elements of S, the only solution of the equation $\sum_{i=1}^{n} c_i s_i = 0$, is $c_1 = \cdots = c_n = 0$. If a subset of X is not linearly independent, it is called *linearly dependent*. If S is linearly independent, one can show that any nonzero $x \in \text{span} S$ admits a unique representation $x = \sum_{i=1}^{n} c_i s_i$, where $s_i \in S$, and c_i's are nonzero numbers, $i = 1, \ldots, n$.

A linearly independent subset S of a linear space X is called a *(Hamel) basis* for X if $X = \text{span} S$. One can show that different bases of X have the same number of members. This number is called the *dimension* of X and is denoted by $\dim X$. If $\dim X$ is finite, X is called a *finite-dimensional* linear space, otherwise, X is called an *infinite-dimensional* linear space.

Example 2.19. Consider the set of polynomials $S = \{1, x, x^2\} \subset \mathbb{P}_2(\mathbb{R})$. Then S is linearly independent, since let $c_0 + c_1 x + c_2 x^2 = 0$, for any $x \in \mathbb{R}$. If at least one of the constant c_i is not zero, we have a polynomial on the left side of the equality, which has at most 2 zeros. But this is a contradiction as the left side must be zero for all $x \in \mathbb{R}$, and therefore, $c_0 = c_1 = c_2 = 0$, and S is linearly independent. As $\operatorname{span} S = \mathbb{P}_2(\mathbb{R})$, we conclude that $\dim \mathbb{P}_2(\mathbb{R}) = 3$. More generally, one can show that $\dim \mathbb{P}_k(\mathbb{R}^n) = \binom{n+k}{k} = \frac{(n+k)!}{k!n!}$.

Example 2.20. Let $\mathbb{P}(\mathbb{R})$ be the space of polynomials of all degrees $k \geq 0$ in $\mathbb{R}$ and let $S = \{1, x, x^2, x^3, \dots\}$. By considering an arbitrary linear combination of finite members of S and following the approach of Example 2.19, it is straightforward to show that S is linearly independent. Since $\operatorname{span} S = \mathbb{P}(\mathbb{R})$, we conclude that $\mathbb{P}(\mathbb{R})$ is an infinite-dimensional linear space. Since $\mathbb{P}(\mathbb{R})$ is a linear subspace of $C^m(\overline{\Omega})$, $C^m(\Omega)$, $L^2(\Omega)$, and $H^1(\Omega)$, we conclude that these spaces are also infinite-dimensional. Moreover, the partly Sobolev classes $H(\operatorname{div}; \Omega)$ and $H(\operatorname{curl}; \Omega)$ are also infinite-dimensional as the space of vector fields with $\mathbb{P}(\mathbb{R})$ components is a subspace of these partly Sobolev classes.

Let $L : X \to Y$ be a linear mapping. The *rank-nullity theorem* states that

$$\dim N(L) + \dim R(L) = \dim X. \tag{2.1}$$

Example 2.21. For the linear mapping $L : \mathbb{R}^3 \to \mathbb{R}^3$ of Example 2.15, we have $\dim N(L) = 2$, and $\dim R(L) = 1$ and therefore, $\dim N(L) + \dim R(L) = 3$.

Example 2.22. Let $M : \mathbb{R}^n \to \mathbb{R}^n$ be a linear mapping. If M is surjective, we have $\dim R(M) = n$, and then (2.1) implies that M is injective as $\dim N(L) = 0$. Conversely, (2.1) implies that if M is injective, then it is also surjective. Thus, a linear mapping $\mathbb{R}^n \to \mathbb{R}^n$ is injective if and only if it is surjective.

Example 2.23. As discussed in Example 2.12, a matrix $M_{m \times n}$ can be considered as a linear mapping $\mathbb{R}^n \to \mathbb{R}^m$. The maximum rank of this linear mapping is m and occurs if it is surjective. In this case, the matrix M is called *full rank*.

2.5   THE MATRIX REPRESENTATION OF LINEAR MAPPINGS AND BILINEAR FORMS

Suppose X and Y are finite-dimensional linear spaces with the bases $B_X = \{x_1, \dots, x_n\}$, and $B_Y = \{y_1, \dots, y_m\}$, respectively. Any linear mapping $L : X \to Y$ can be represented by an $m \times n$ matrix as follows: Let $L(x_i) = \sum_{j=1}^{m} l_{ji} y_j$, where l_{ji} is the j-th component of $L(x_i)$ in the basis B_Y. Since any $x \in X$ can be uniquely written as $x = \sum_{i=1}^{n} a_i x_i$, we can write

$$L(x) = L\left(\sum_{i=1}^{n} a_i x_i\right) = \sum_{i=1}^{n} a_i L(x_i) = \sum_{i=1}^{n} a_i \cdot \left(\sum_{j=1}^{m} l_{ji} y_j\right) = \sum_{j=1}^{m} \left(\sum_{i=1}^{n} l_{ji} a_i\right) y_j.$$

In the basis B_Y, $L(x) \in Y$ admits a unique representation $L(x) = \sum_{j=1}^{m} b_j y_j$, and therefore, the above relation implies that $b_j = \sum_{i=1}^{n} l_{ji} a_i$, $j = 1, \ldots, m$. This equation can be stated in the matrix form $\mathbf{b} = [L] \cdot \mathbf{a}$, where

$$
\mathbf{b}_{m \times 1} = \begin{bmatrix} b_1 \\ b_2 \\ \vdots \\ b_m \end{bmatrix}, \ [L]_{m \times n} = \begin{bmatrix} l_{11} & l_{12} & \cdots & l_{1n} \\ l_{21} & l_{22} & \cdots & l_{2n} \\ \vdots & \vdots & \ddots & \vdots \\ l_{m1} & l_{m2} & \cdots & l_{mn} \end{bmatrix}, \ \mathbf{a}_{n \times 1} = \begin{bmatrix} a_1 \\ a_2 \\ \vdots \\ a_n \end{bmatrix}. \tag{2.2}
$$

Notice that $\mathbf{a}$ and $\mathbf{b}$ respectively contain the components of x in the basis B_X and the components of $L(x)$ in the basis B_Y. The matrix $[L]$ is *a matrix representation of L*. This matrix representation is not unique and depends on the chosen bases for X and Y.

Example 2.24. Consider the linear mapping $D : \mathbb{P}_3(\mathbb{R}) \to \mathbb{P}_2(\mathbb{R})$, $D(p) = 2 \frac{dp(x)}{dx}$. We choose the basis $B_1 = \{1, x, x^2, x^3\}$ for $\mathbb{P}_3(\mathbb{R})$ and the basis $B_2 = \{1, x, x^2\}$ for $\mathbb{P}_2(\mathbb{R})$. Since $D(1) = 0$, $D(x) = 2$, $D(x^2) = 4x$, and $D(x^3) = 6x^2$, the matrix representation of L in the bases B_1 and B_2 reads

$$
[D] = \begin{bmatrix} 0 & 2 & 0 & 0 \\ 0 & 0 & 4 & 0 \\ 0 & 0 & 0 & 6 \end{bmatrix}.
$$

Different bases for $\mathbb{P}_3(\mathbb{R})$ and $\mathbb{P}_2(\mathbb{R})$ lead to different representations for D, see Exercise 2.15.

Similarly, one can also write matrix representation of bilinear forms. More specifically, let $B : X \times Y \to \mathbb{R}$ be a bilinear form. Using the bases B_X and B_Y, one can write

$$
B(x, y) = B \left(\sum_{i=1}^{n} a_i x_i, \sum_{j=1}^{m} b_j y_j \right) = \sum_{j=1}^{m} \sum_{i=1}^{n} b_j B(x_i, y_j) a_i,
$$

which is equivalent to $B(x, y) = \mathbf{b}^T \cdot [B] \cdot \mathbf{a}$, where the vector $\mathbf{a}$ and $\mathbf{b}$ are introduced in (2.2) and the matrix $[B]_{m \times n}$ is the matrix representation of the bilinear form in the bases B_X and B_Y given by

$$
[B]_{m \times n} = \begin{bmatrix} B(x_1, y_1) & B(x_2, y_1) & \cdots & B(x_n, y_1) \\ B(x_1, y_2) & B(x_2, y_2) & \cdots & B(x_n, y_2) \\ \vdots & \vdots & \ddots & \vdots \\ B(x_1, y_m) & B(x_2, y_m) & \cdots & B(x_n, y_m) \end{bmatrix}.
$$

Notice that the ij-th component of $[B]$ is $B(x_j, y_i)$.

Example 2.25. Let $\Omega = (-1, 1) \subset \mathbb{R}$ and consider the bilinear form $B : \mathbb{P}_2(\Omega) \times \mathbb{P}_2(\Omega) \to \mathbb{R}$, $B(f, g) = \int_{\Omega} fg$. To write the matrix representation of B in the basis

$\{1,x,x^2\}$, we note that

$$B(1,1) = \int_{-1}^{1} dx = 2, \ \ B(1,x) = \int_{-1}^{1} x\,dx = 0, \ \ B(1,x^2) = B(x,x) = \int_{-1}^{1} x^2\,dx = \frac{2}{3},$$

$$B(x,x^2) = \int_{-1}^{1} x^3\,dx = 0, \ \ B(x^2,x^2) = \int_{-1}^{1} x^4\,dx = \frac{2}{5},$$

and since B is symmetric, we conclude that

$$[B] = \begin{bmatrix} 2 & 0 & \frac{2}{3} \\ 0 & \frac{2}{3} & 0 \\ \frac{2}{3} & 0 & \frac{2}{5} \end{bmatrix}.$$

2.6 NORMED LINEAR SPACES

The notion of *norm* generalizes the notion of length to linear spaces. More specifically, a norm on a linear space X is a real-valued function $\|\cdot\| : X \to \mathbb{R}$ such that

1. $\|x\| \geq 0$, for all $x \in X$ (*positivity*);
2. $\|x+y\| \leq \|x\| + \|y\|$, for any $x,y \in X$ (*the triangle inequality*);
3. $\|\alpha x\| = |\alpha|\|x\|$, for any $x \in X$ and $\alpha \in \mathbb{R}$;
4. $\|x\| = 0$, if and only if $x = 0$ (*positive definiteness*).

A *normed linear space* $(X, \|\cdot\|)$ is a linear space X together with a norm $\|\cdot\|$ on X. For any $x \in X$, $\|x\|$ is called *the norm of x*.

Example 2.26. A simple example of normed linear spaces is $(\mathbb{R}^n, \|\cdot\|)$, where $\|\cdot\|$ is the standard norm of $\mathbb{R}^n$, that is, for any $x = (x_1,\ldots,x_n) \in \mathbb{R}^n$, we have $\|x\| = (x_1^2 + \cdots + x_n^2)^{1/2}$.

Example 2.27. In this example and the next one, we introduce some normed linear spaces which are frequently encountered in the approximation of solutions of differential equations by using finite element methods. The L^2-norm $\|\cdot\|_2$ on the Lebesgue space $L^2(\Omega)$ of Example 2.7 is given by

$$\|f\|_2 = \left(\int_\Omega |f|^2 \right)^{\frac{1}{2}}, \ f \in L^2(\Omega).$$

The L^2-norm on $[L^2(\Omega)]^n$ is componentwise defined, that is, for any $v = (v_1,\ldots,v_n) \in [L^2(\Omega)]^n$, we have

$$\|v\|_2 = \left(\|v_1\|_2^2 + \cdots + \|v_n\|_2^2 \right)^{\frac{1}{2}}.$$

For any $f \in H^1(\Omega)$, the H^1-norm $\|f\|_{1,2}$ on the Sobolev space $H^1(\Omega)$ of Example 2.8 is given by

$$\|f\|_{1,2} = \left(\|f\|_2^2 + \|\partial_{x_1} f\|_2^2 + \cdots + \|\partial_{x_n} f\|_2^2 \right)^{\frac{1}{2}}.$$

Finally, for any $v = (v_1, \ldots, v_n) \in [H^1(\Omega)]^n$, the H^1-norm on $[H^1(\Omega)]^n$ is componentwise defined as

$$\|v\|_{1,2} = \left(\|v_1\|_{1,2}^2 + \cdots + \|v_n\|_{1,2}^2 \right)^{\frac{1}{2}}.$$

Example 2.28. By using the Partly Sobolev classes of Example 2.9, one can define the normed linear spaces $(H(\mathrm{div}; \Omega), \|\cdot\|_d)$ and $(H(\mathrm{curl}; \Omega), \|\cdot\|_c)$, where the associated norm of a vector field v belonging to these spaces is respectively defined as

$$\|v\|_d = \left(\|v\|_2^2 + \|\mathrm{div}\, v\|_2^2 \right)^{\frac{1}{2}}, \text{ and } \|v\|_c = \left(\|v\|_2^2 + \|\mathrm{curl}\, v\|_2^2 \right)^{\frac{1}{2}}.$$

The standard norm of the Euclidean space $\mathbb{R}^n$ enables us to measure the distance between different points $x, y \in \mathbb{R}^n$ by using $\|x - y\|$. Similarly, we may measure the "distance" between two arbitrary points $x, y \in X$ of a normed linear space $(X, \|\cdot\|)$ by using $\|x - y\|$.

Two functions $f_1, f_2 \in L^2(\Omega)$ are "close" to each other in $L^2(\Omega)$ if $\|f_1 - f_2\|_2$ is small. Similarly, $f_1, f_2 \in H^1(\Omega)$ are close to each other in $H^1(\Omega)$ if $\|f_1 - f_2\|_{1,2}$ is small. Suppose f_1 and f_2 belong to both $L^2(\Omega)$ and $H^1(\Omega)$. Then, f_1 and f_2 may be close in $L^2(\Omega)$ but not in $H^1(\Omega)$ since the smallness of $\|f_1 - f_2\|_2$ suggests that values of f_1 and f_2 are close, however, the smallness of $\|f_1 - f_2\|_{1,2}$ implies that not only values but also derivatives of f_1 and f_2 are close, see Figure 2.1. In numerical analysis, norms are employed to measure the difference between *approximate* solutions and the *exact* solutions of problems.

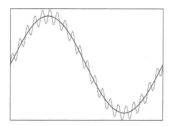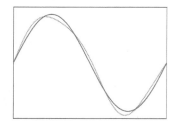

Figure 2.1 The two functions of the left panel are "close" in the L^2-norm but not in the H^1-norm while the functions on the right panel are close in both norms.

2.7 FUNCTIONALS AND DUAL SPACES

Let X be a linear space. Any real-valued mapping $J : X \to \mathbb{R}$ is called a *functional*. If J is linear, it is called a *linear functional*. Some physical quantities such as energy can be expressed as functionals.

Example 2.29. The mapping $J : C^0(\overline{\Omega}) \to \mathbb{R}$, $J(u) = \int_\Omega u^2$ is a functional. For any $v \in C^0(\overline{\Omega})$, the mapping $f : C^0(\overline{\Omega}) \to \mathbb{R}$, $f(u) = \int_\Omega uv$, is a linear functional.

Let X be a normed linear space. The space of all continuous linear functionals over X is called *the dual space* of X and is denoted by X' or $L(X; \mathbb{R})$. By using the pointwise operations, it is easy to show that X' is a linear space. It is customary to denote the value of $f \in X'$ at $x \in X$ using *the duality brackets* as $f(x) = \langle f, x \rangle$.

Example 2.30 (Dual bases). Let X be a finite-dimensional normed linear space. Then, X' is equal to the space of all linear functionals. Suppose $B_X = \{x_1, \ldots, x_n\}$ is a basis for X and consider the set of linear functionals $B_{X'} = \{f_1, \ldots, f_n\}$, with $f_i(x_j) = \delta_{ij}$, where *the Kronecker delta* δ_{ij} is 1, if $i = j$, and 0 otherwise. It is not hard to show that $B_{X'}$ is a basis for X': Let $g \in X'$ be an arbitrary functional. Since B_X is a basis for X, any $x \in X$ admits a unique representation $x = \sum_{i=1}^{n} a_i x_i$. We can write

$$g(x) = g\left(\sum_{i=1}^{n} a_i x_i \right) = \sum_{i=1}^{n} g(x_i) a_i = \sum_{i=1}^{n} g(x_i) \left(\sum_{j=1}^{n} a_j \delta_{ij} \right) = \sum_{i=1}^{n} g(x_i) \left(\sum_{j=1}^{n} a_j f_i(x_j) \right)$$

$$= \sum_{i=1}^{n} g(x_i) f_i \left(\sum_{j=1}^{n} a_j x_j \right) = \sum_{i=1}^{n} g(x_i) f_i(x),$$

which implies that all members of X' can be written as a linear combination of the elements of $B_{X'}$, that is, $X' = \operatorname{span} B_{X'}$. On the other hand, let $\sum_{i=1}^{n} c_i f_i(x) = 0$, for some constants $c_1, \ldots, c_n$, and any $x \in X$. In particular, at $x = x_j$, we have $0 = \sum_{i=1}^{n} c_i f_i(x_j) = \sum_{i=1}^{n} c_i \delta_{ij} = c_j$, $j = 1, \ldots, n$. Thus, $B_{X'}$ is also linearly independent. Hence, we conclude that $B_{X'}$ is a basis for X' and $\dim X = \dim X'$. The basis B_X is said to be *the dual basis* of $B_{X'}$. In the next chapter, we will see that finite elements can be defined by using dual bases.

Example 2.31. Let $X = \mathbb{P}_1(K)$, with $K = [0, 1]$, and consider the set $B_{X'} = \{f_1, f_2\} \subset X'$, where $f_1(p) = p(0)$, and $f_2(p) = p(1)$. We show that $B_{X'}$ is a basis for X'. Let $c_1 f_1(p) + c_2 f_2(p) = 0$, for all $p \in \mathbb{P}_1(K)$. In particular, for $p_1(x) = 1 - x$, and $p_2(x) = x$, we obtain $0 = c_1 f_1(p_1) + c_2 f_2(p_1) = c_1 p_1(0) + c_2 p_1(1) = c_1$, and $0 = c_1 f_1(p_2) + c_2 f_2(p_2) = c_1 p_2(0) + c_2 p_2(1) = c_2$. Therefore, $B_{X'}$ is linearly independent and since $\dim X' = \dim X = 2$, we conclude that $B_{X'}$ is a basis for X'. Moreover, since $f_j(p_i) = \delta_{ij}$, $B_X = \{p_1, p_2\}$ is the dual basis of $B_{X'}$.

2.8   GREEN'S FORMULAS

Let $\Omega \subset \mathbb{R}^m$ and let $\mathbf{n} = (n_1, \ldots, n_m)$ be *the outward unit normal vector field* at the boundary $\partial \Omega$. *The fundamental Green's formula* states that if $v \in C^1(\overline{\Omega})$, then

$$\int_{\Omega} \partial_{x_i} v = \int_{\partial \Omega} v n_i, \ i = 1, \ldots, m. \tag{2.3}$$

By using this formula, one can write other Green's formulas. For example, given a function v and a vector field $\mathbf{u} = (u_1, \ldots, u_m)$, one can show that

$$\int_{\Omega} \mathbf{u} \cdot \nabla v = - \int_{\Omega} v \operatorname{div} \mathbf{u} + \int_{\partial \Omega} (v\mathbf{u}) \cdot \mathbf{n}. \tag{2.4}$$

where "$\cdot$" is the standard *inner product* of $\mathbb{R}^m$ given by $\mathbf{u} \cdot \mathbf{v} = \sum_{i=1}^{m} u_i v_i$, and $\nabla v = (\partial_{x_1} v, \ldots, \partial_{x_m} v)$ is the *gradient* of v.

Another Green's formula for *the Laplacian* $\Delta u = \sum_{i=1}^{m} \partial_{x_i x_i} u = \mathrm{div}(\nabla u)$, can be written as follows: Let

$$\partial_n v = \boldsymbol{n} \cdot \nabla v = \sum_{i=1}^{m} n_i \partial_{x_i} v,$$

be *the normal derivative* of v at the boundary $\partial \Omega$. Then, for any functions u and v, we have

$$\int_{\Omega} \nabla u \cdot \nabla v = - \int_{\Omega} v \Delta u + \int_{\partial \Omega} v \partial_n u. \tag{2.5}$$

The above Green's formulas can be extended to Sobolev spaces based on the following results:

1. If $v, u \in L^2(\Omega)$ then $\int_{\Omega} uv$ exists;
2. If $u \in H^1(\Omega)$, then $u|_{\partial \Omega} \in L^2(\partial \Omega)$ and $\int_{\partial \Omega} u$ exists, however, this conclusion generally does not hold if u only belongs to $L^2(\Omega)$;
3. If $v \in H^1(\Omega)$ and $\boldsymbol{u} \in H(\mathrm{div}; \Omega)$, then $\boldsymbol{u} \cdot \boldsymbol{n}$ belongs to $L^2(\partial \Omega)$ and $\int_{\partial \Omega} (v\boldsymbol{u}) \cdot \boldsymbol{n}$ exists.

By using these results, one can show that Green's formula (2.4) holds either if $\boldsymbol{u} \in [H^1(\Omega)]^n$ and $v \in H^1(\Omega)$, or if $\boldsymbol{u} \in H(\mathrm{div}; \Omega)$ and $v \in H^1(\Omega)$. Similar results hold for (2.5) as well. As will be discussed in the following chapters, Green's formulas in terms of Sobolev spaces are extensively used in the finite element method to write weak forms of governing equations.

EXERCISES

Exercise 2.1. Suppose $B \subset \mathbb{R}$ is a non-empty set. Show that: (i) If $\max B$ exists, then $\max B = \sup B$, and (ii) if $\min B$ exists, then $\min B = \inf B$.

Exercise 2.2. Let $\Omega \subset \mathbb{R}^n$ be an open subset. Is $\mathbb{P}_k(\Omega)$ a linear subspace of $L^2(\Omega)$, $C^m(\Omega)$, and $C^m(\overline{\Omega})$?

Exercise 2.3. Let $\Omega = (0,1) \times (0,1)$ be *the unit square* and let $f(x,y) = (x+y)^b$. Find values of b such that: (i) $f \in L^2(\Omega)$; (ii) $f \in H^1(\Omega)$; and (iii) $f \in H^2(\Omega)$.

Exercise 2.4. Show that affine mappings are not linear.

Exercise 2.5. Let ∇f denote the gradient of $f \in C^1(\overline{\Omega})$ and let "$\cdot$" denote the standard inner product of $\mathbb{R}^n$. Show that $B : C^1(\overline{\Omega}) \times C^1(\overline{\Omega}) \to \mathbb{R}$, $B(f,g) = \int_{\Omega} \nabla f \cdot \nabla g$, is a symmetric bilinear form.

Exercise 2.6. Let X and Y be linear spaces. Then, $Z = X \times Y$ is also a linear space with the addition $(x_1, y_1) + (x_2, y_2) = (x_1 + x_2, y_1 + y_2)$, for all $(x_i, y_i) \in Z$, $i = 1, 2$, and the scalar multiplication $c(x,y) = (cx, cy)$, for all $c \in \mathbb{R}$. Suppose $B : X \times Y \to \mathbb{R}$ is a bilinear form. Show that B as the mapping $B : Z \to \mathbb{R}$ is not a linear mapping.

Exercise 2.7. Let $L : X \to Y$ be a linear mapping. Show that: (i) $\ker L$ is a linear subspace of X; (ii) $R(L)$ is a linear subspace of Y; and (iii) the linear mapping L is injective if and only if $\ker L = \{0\}$.

Exercise 2.8. Show that if a linear mapping $L : X \to Y$ is invertible, then its inverse $L^{-1} : Y \to X$ is also a linear mapping.

Exercise 2.9. Let S be a subset of a linear space X. Show that $\operatorname{span} S$ is a linear subspace of X.

Exercise 2.10. By writing a basis, show that: (i) $\dim \mathbb{P}_2(\mathbb{R}^2) = 6$; (ii) $\dim \mathbb{P}_2(\mathbb{R}^3) = 10$; (iii) $\dim \mathbb{P}_3(\mathbb{R}^2) = 10$; and (iv) $\dim \mathbb{P}_3(\mathbb{R}^3) = 20$.

Exercise 2.11. Let $\mathbb{Q}_2(\mathbb{R}^2)$ be the space of all polynomials of degree less than or equal to 2 with respect to each one of the two variables x and y. By deriving a basis for this linear space show that $\dim \mathbb{Q}_2(\mathbb{R}^2) = 9$. Also show that $\mathbb{Q}_2(\mathbb{R}^2)$ is a linear subspace of $\mathbb{P}_k(\mathbb{R}^2)$, $k \geq 4$.

Exercise 2.12. By generalizing the polynomial space of Exercise 2.11, define $\mathbb{Q}_k(\mathbb{R}^n)$ and show that $\dim \mathbb{Q}_k(\mathbb{R}^n) = (k+1)^n$. This space is useful for defining finite elements over rectangular elements.

Exercise 2.13. Let $L : \mathbb{R}^n \to \mathbb{R}^m$ be a linear mapping. What can we say about the injectivity, the surjectivity, and the bijectivity of L if (i) $n > m$; (ii) $n = m$; and (iii) $n < m$? Justify your answers by using the rank-nullity theorem and give an example for each case.

Exercise 2.14. Suppose a non-singular matrix $M_{k \times k}$ (that is, M as mapping $\mathbb{R}^k \to \mathbb{R}^k$ is bijective) is of the form

$$M_{k \times k} = \left[\begin{array}{c:c} A_{n \times n} & B_{n \times l} \\ \hdashline C_{l \times n} & \mathbf{0} \end{array} \right],$$

with $k = n + l$. Show that the inequality $l \leq n$ must hold. In Chapters 4 and 5, we will use similar results to study the stability of mixed finite element methods. (*Hint.* Use the fact that the bijectivity of M implies the surjectivity of C and then employ Exercise 2.13.)

Exercise 2.15. Show that $\tilde{B}_1 = \{1 + 2x, x, x^2 + x^3, x^3\}$ and $\tilde{B}_2 = \{2, x+1, x^2\}$ are bases for $\mathbb{P}_3(\mathbb{R})$ and $\mathbb{P}_2(\mathbb{R})$, respectively. Obtain the matrix representation of the linear mapping of Example 2.24 in these bases.

Exercise 2.16. Let Ω be the open interval $(-1, 1)$ and consider the bilinear form $B : \mathbb{P}_3(\Omega) \times \mathbb{P}_3(\Omega) \to \mathbb{R}$, $B(f, g) = \int_\Omega \nabla f \cdot \nabla g$. Derive the matrix representation of B in the basis $\{1, x, x^2, x^3\}$ of $\mathbb{P}_3(\Omega)$.

Exercise 2.17. Let $x = (x_1, \ldots, x_n) \in \mathbb{R}^n$. Show that $\|x\| = |x_1| + \cdots + |x_n|$ is a norm on $\mathbb{R}^n$. (*Hint.* To prove the triangle inequality, you can use the triangle inequality of real numbers, that is, $|a + b| \leq |a| + |b|$ for any $a, b \in \mathbb{R}$.)

Exercise 2.18. Let $X = \mathbb{P}_1(K)$, with $K = [a_1, a_2]$. Show that $B_{X'} = \{f_1, f_2\}$, where $f_i(p) = p(a_i)$, $i = 1, 2$, is a basis for X'. Obtain the dual basis of $B_{X'}$. In Chapter 3, we will see that $B_{X'}$ contains the degrees of freedom of a Lagrange finite element of degree 1.

Exercise 2.19. Let $X = \mathbb{P}_2(K)$, with $K = [a_1, a_2]$. Show that $B_{X'} = \{f_1, f_2, f_3\}$, where $f_i(p) = p(a_i)$, $i = 1, 2, 3$, and $a_3 = (a_1 + a_2)/2$, is a basis for X' and obtain the dual basis of $B_{X'}$. In Chapter 3, we will see that $B_{X'}$ contains the degrees of freedom of a Lagrange finite element of degree 2.

Exercise 2.20. Derive Green's formula (2.4) by using the fundamental Green's formula (2.3).

Exercise 2.21. Derive Green's formula (2.5) by using Green's formula (2.4) and the relation $\Delta u = \text{div}(\nabla u)$.

COMMENTS AND REFERENCES

Discussions on real numbers and subsets of the Euclidean space and their properties are available in standard real analysis textbooks such as [3]. A good introduction to linear spaces and linear operators in engineering and science can be found in [11]. A classical text for standard Sobolev spaces is [1]. An introductory discussion on partly Sobolev classes $H(\text{div}; \Omega)$ and $H(\text{curl}; \Omega)$ is mentioned in [4, Chapter 2]. A brief discussion of Green's formulas in the context of solving partial differential equations by using the finite element method is given in [5, Section 1.2].

3 Finite Element Interpolation

The convergence of finite element solutions to the true solution of physical models is closely related to the general capability of finite elements to approximate arbitrary functions. This process of approximating functions with finite elements is called *finite element interpolation*. In this chapter, we introduce several finite elements and discuss how they can be employed to interpolate functions and vector fields.

3.1 1D FINITE ELEMENT INTERPOLATION

To motivate the concept of finite elements and the main ideas of function interpolation using finite elements, we begin this chapter by studying piecewise linear interpolations of functions defined on an open interval of $\mathbb{R}$.

3.1.1 THE GLOBAL LEVEL

Consider the unit interval $\Omega = (0,1)$. Given a function $f : [0,1] \to \mathbb{R}$, a simple approach for *interpolating* (also called *approximating*) f is to use piecewise linear functions such as the one shown in Figure 3.1. The first step for obtaining such interpolating functions (also called approximating functions) using finite elements is to establish a *mesh* T_h on Ω. To this end, we consider a set of points $\{x_0, x_1, \ldots, x_N\}$ such that $0 = x_0 < x_1 < \cdots < x_N = 1$. The mesh T_h induced by these points is the collection of subintervals $\{K_0, \ldots, K_{N-1}\}$ with $K_i = [x_i, x_{i+1}]$. The index h of T_h refers to the maximum length h of subintervals, that is, $h = \max\{h_0, \ldots, h_{N-1}\}$, where h_i is the length of K_i. The points x_i and the subintervals K_i are then called the *vertices* and the *elements* (or *cells*) of T_h, respectively.

Figure 3.1 suggests that by using the values $\{f(x_0), \ldots, f(x_N)\}$ of f at the vertices of T_h, we can uniquely specify a piecewise linear function $I_h f$ with the following properties: (i) $I_h f(x_i) = f(x_i)$, for all vertices x_i, and (ii) $f|_{K_i} \in \mathbb{P}_1(K_i)$ for all elements K_i. The function $I_h f$ is called *the Lagrange interpolant of f of type* (1). The values $\{f(x_0), \ldots, f(x_N)\}$ that uniquely determine $I_h f$ are called *global degrees of freedom*. Here the adjective *global* means these degrees of freedom are assigned to the whole mesh T_h.

Notice that $I_h f$ belongs to the set $\mathbb{P}_1(T_h)$ that includes all continuous, piecewise linear functions over T_h, that is

$$\mathbb{P}_1(T_h) = \{v \in C^0(\overline{\Omega}) : v|_{K_i} \in \mathbb{P}_1(K_i), \ i = 0, \ldots, N-1\}. \tag{3.1}$$

This set is then called an *approximation space* as it includes the approximating function $I_h f$. One can use the fact that $\mathbb{P}_1(T_h)$ is a linear space to write $I_h f$ in terms of a

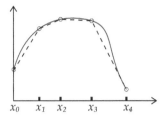

Figure 3.1 A function (the solid line) and its piecewise linear interpolant (the dashed line) associated to the vertices $\{x_0, \ldots, x_4\}$.

basis of $\mathbb{P}_1(T_h)$. More specifically, consider the *hat* functions $\psi_i \in \mathbb{P}_1(T_h)$, $i = 0, \ldots, N$, given by

$$\psi_i(x) = \begin{cases} \frac{1}{h_{i-1}}(x - x_{i-1}), & \text{if } x \in K_{i-1}, \\ \frac{1}{h_i}(x_{i+1} - x), & \text{if } x \in K_i, \\ 0, & \text{otherwise.} \end{cases} \tag{3.2}$$

For the cases $i = 0, N$, straightforward modifications of this definition are needed as shown in Figure 3.2. We have

$$\psi_j(x_i) = \delta_{ij}, \ i, j = 0, \ldots, N. \tag{3.3}$$

Also the hat functions are non-zero only on a few elements.

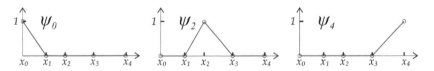

Figure 3.2 The basis functions ψ_0, ψ_2, and ψ_4, which are respectively associated to the vertices x_0, x_2, and x_4.

One can show that $\{\psi_0, \psi_1, \ldots, \psi_N\}$ is a basis for $\mathbb{P}_1(T_h)$, and therefore, $\dim \mathbb{P}_1(T_h) = N + 1$. Thus, any $v \in \mathbb{P}_1(T_h)$ can be uniquely written as $v = \sum_{j=0}^{N} c_j \psi_j$. Using (3.3), we can write $v(x_i) = \sum_{j=0}^{N} c_j \psi_j(x_i) = c_i$, and therefore,

$$v(x) = \sum_{i=0}^{N} v(x_i) \psi_i(x).$$

For any continuous function $f \in C^0(\overline{\Omega})$, since $I_h f \in \mathbb{P}_1(T_h)$ and $I_h f(x_i) = f(x_i)$, we conclude that

$$I_h f = \sum_{i=0}^{N} f(x_i) \psi_i.$$

Thus, the mesh T_h induces *the interpolation operator* $I_h : C^0(\overline{\Omega}) \to \mathbb{P}_1(T_h)$ that assigns the piecewise linear interpolant $I_h f$ to any continuous function f. In summary, the basis functions ψ_i on T_h allow us to approximate functions.

The basis functions ψ_i are called *the global shape functions* associated to the global degree of freedom at the vertex x_i. The relation (3.3) implies that the value of ψ_i is 1 at the corresponding vertex and zero at other vertices.

A crucial question regarding the Lagrange interpolant $I_h f$ is its ability to approximate f in the following sense: Suppose $\{T_{h_1}, T_{h_2}, \dots\}$ is a sequence of meshes of Ω such that $h_1 > h_2 > h_3 > \cdots$, and $h_i \to 0$ as $i \to \infty$, that is, $T_{h_{i+1}}$ is more refined than T_{h_i}. Let $I_h f$ be the associated Lagrange interpolants of f with $h = h_1, h_2, \dots$, see Figure 3.3. We want to know if $I_h f$ "converges" to f as $h \to 0$.

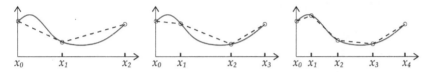

Figure 3.3 By increasing the refinement level of meshes, Lagrange interpolants (the dashed lines) provide more accurate approximations of a function (the solid lines).

The normed linear spaces $\big(L^2(\Omega), \|\cdot\|_2\big)$ and $\big(H^1(\Omega), \|\cdot\|_{1,2}\big)$ introduced in Example 2.27 provide a suitable framework to answer the above question. Recall that norms can measure the distance between members of linear spaces. Since $\mathbb{P}_1(T_h) \subset H^1(\Omega) \subset L^2(\Omega)$, we can interpret the convergence of $I_h f$ to f as $h \to 0$ as follows: *Does $\|f - I_h f\| \to 0$ as $h \to 0$, where $\|\cdot\|$ can be either $\|\cdot\|_2$ or $\|\cdot\|_{1,2}$?*

We can use the Python based package FEniCS to examine the above question. A brief discussion on FEniCS installation and a short introduction to Python are provided in the appendices. Further references for FEniCS and Python are mentioned at the end of this chapter. The following code calculates the errors $\|f - I_h f\|_2$ and $\|f - I_h f\|_{1,2}$ for $f(x) = \sin(\pi x)$ by using 4 meshes containing $3, 6, 9,$ and 12 elements of the same size.

```
from dolfin import *

# defining the function
f = Expression('sin(pi*x[0])', degree = 5)

# number of divisions of meshes
Divisions = [3, 6, 9, 12]

# computing interpolants and errors
h = []                          # max element sizes
error_L2, error_H1 = [], []    # initializing errors
```

```
for n in Divisions:
    # defining the mesh
    mesh = UnitIntervalMesh(n)

    # defining the approximation space
    CGE = FiniteElement("Lagrange", mesh.ufl_cell(), 1)
    Z = FunctionSpace(mesh, CGE)

    # interpolating f
    Interpolant = interpolate(f, Z)

    # max element size of the mesh
    h.append(mesh.hmax())

    # calculating errors
    error_L2.append(errornorm(f, Interpolant, norm_type="L2"))
    error_H1.append(errornorm(f, Interpolant, norm_type="H1"))
```

Let us study the above program. The first line

```
from dolfin import *
```

imports FEniCS classes such as `Expression` and `UnitIntervalMesh` from its DOLFIN library. Alternatively, one may also use

```
from fenics import *
```

FEniCS programs usually start with one of the above lines. The statement

```
f = Expression('sin(pi*x[0])', degree = 5)
```

defines the function $f(x) = \sin(\pi x)$. The formula is expressed as a string with C++ syntax. In 1D, `x[0]` refers to the x coordinate. In 2D and 3D, the coordinates (x, y) and (x, y, z) are denoted by `(x[0],x[1])` and `(x[0],x[1],x[2])`, respectively. Expressions are interpolated in FEniCS and the degree of polynomials that will be used for this purpose are specified by the argument `degree = 5`. The list

```
Divisions = [3, 6, 9, 12]
```

specifies the number of divisions of meshes. In the for loop, interpolants and interpolation errors are calculated for each mesh. A mesh of the unit interval $[0, 1]$ containing n elements of the same size is defined by

```
mesh = UnitIntervalMesh(n)
```

It is easy to obtain mesh attributes. For example, `mesh.ufl_cell()` and `mesh.hmax()` respectively denote the shape of the elements and the maximum element size of `mesh`. The approximation space $\mathbb{P}_1(T_h)$, denoted by Z, is then specified as

```
CGE = FiniteElement("Lagrange", mesh.ufl_cell(), 1)
Z = FunctionSpace(mesh, CGE)
```

As we will discuss in the next section, this approximation space can be constructed by using Lagrange finite elements. The first line assigns a degree 1 Lagrange element to CGE, where as we mentioned earlier, `mesh.ufl_cell()` denotes the shape of the elements of `mesh`. The approximation space Z is then defined by using the mesh and the finite element CGE. The Lagrange interpolant is simply obtained by

```
Interpolant = interpolate(f, Z)
```

Having the function f and `Interpolant`, the L^2 and H^1 errors can be computed using the statement

```
errornorm(f, Interpolant, norm_type)
```

where `norm_type` is a string specifying the type of error. Figure 3.4 shows the output of the above program: The left panel depicts the interpolants for different meshes and the right panel shows the L^2- and the H^1-errors versus h.

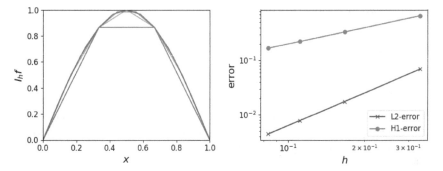

Figure 3.4   Lagrange Interpolants of type (1) for $f(x) = sin(\pi x)$ on $[0, 1]$: (Left) Interpolants for meshes with 3, 6, 9, and 12 uniform elements; (Right) L^2- and H^1-errors versus h.

It is possible to show that when f is twice differentiable, we have

$$\|f - I_h f\|_2 \le Ch^2, \tag{3.4}$$

$$\|f - I_h f\|_{1,2} \le \tilde{C}h, \tag{3.5}$$

where C and $\tilde{C}$ do not dependent on h. *These inequalities show that* $\|f - I_h f\| \to 0$ *as* $h \to 0$, *in both the* L^2- *and the* H^1-*norms.* The power of h in the inequalities (3.4) and (3.5) is called *the convergence rate* of the interpolations. Clearly, higher convergence rates imply faster convergence as $h \to 0$. Thus, if f is twice differentiable, then the convergence rate is 2 in the L^2-norm and 1 in the H^1-norm.

To estimate the convergence rates by using the above FEniCS program, we can proceed as follows: Suppose error $= Ch^r$, where C is a constant, and let error$_i = Ch_i^r$

and error$_{i-1} = Ch_{i-1}^r$ be respectively the errors associated to the mesh h_i and the mesh h_{i-1}. By dividing the expressions for the errors and solving for r, one obtains

$$r = \frac{\ln(\text{error}_i/\text{error}_{i-1})}{\ln(h_i/h_{i-1})}. \tag{3.6}$$

Notice that the above relation simply implies that *the convergence rate is the slope of the log-log diagram of error versus h*. By using the outputs h, error_L2, and error_H1 of the above program, we can implement the formula (3.6) as follows:

```
from math import log as ln  # log is a dolfin name too
rate_L2 = ln(error_L2[-1]/error_L2[-2])/ln(h[-1]/h[-2])
rate_H1 = ln(error_H1[-1]/error_H1[-2])/ln(h[-1]/h[-2])
print(' L2-convergence rate = %.2f\n H1-convergence rate = %.2f'
      % (rate_L2,rate_H1))
```

For calculating the convergence rates, we used the last two data points that correspond to the meshes with 9 and 12 elements. This program yields the output

```
L2-convergence rate = 2.00
H1-convergence rate = 1.00
```

This output is consistent with the inequalities (3.4) and (3.5). Also notice that the above convergence rates are the slope of the lines shown in the right panel of Figure 3.4.

If f is only one time differentiable, then the convergence rate of $I_h f$ to f is 1 in the L^2-norm and less than 1 in the H^1-norm. *Therefore, the convergence rate depends on the smoothness of f*. The convergence rate 2 in the L^2-norm for the Lagrange interpolant of type (1) is called *the optimal convergence rate* for this type of interpolants in the sense that it is the highest convergence rate due to (3.4).

3.1.2 THE LOCAL LEVEL

Let $T_h = \{K_0, \ldots, K_{N-1}\}$ be a mesh of the unit interval $\Omega = (0,1)$ as discussed in the previous section and suppose $I_h f$ is the Lagrange interpolant of f induced by T_h. An interesting feature of $I_h f$ is that it can be constructed *locally* by using linear interpolants induced by each element. More specifically, suppose $I_i(f|_{K_i})$ is the Lagrange interpolant of the restriction $f|_{K_i}$ of f on the element $K_i = [x_i, x_{i+1}]$, see Figure 3.5. The interpolant $I_i(f|_{K_i})$ is called a *local* interpolant of f in the sense that it is defined only on the element K_i of the mesh T_h. We have

$$\left(I_h f\right)|_{K_i} = I_i(f|_{K_i}).$$

Thus, as shown in Figure 3.5, the interpolant $I_h f$ over T_h can be constructed by assembling all local interpolants $I_i(f|_{K_i})$ associated to the elements K_i of T_h.

The local interpolant $I_i(f|_{K_i})$ belongs to the *local approximation space* $\mathbb{P}_1(K_i)$ that consists of all polynomials of degree ≤ 1 on K_i. Notice that $I_i(f|_{K_i})$ can be

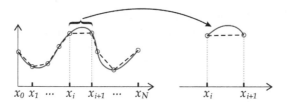

Figure 3.5 Left: A function f (the solid line) and its Lagrange interpolant $I_h f$ (the dashed line) induced by $\{x_0, \ldots, x_N\}$; Right: The restriction of f to $K_i = [x_i, x_{i+1}]$ (the solid line) and the local interpolant $I_i(f|_{K_i})$ on K_i (the dashed line).

uniquely determined by the values $\{f(x_i), f(x_{i+1})\}$. These values at the end points of elements that uniquely specify a member of $\mathbb{P}_1(K_i)$ are called *local degrees of freedom* in $\mathbb{P}_1(K_i)$.

Consider the degree 1 polynomials $\theta_{i,0}$ and $\theta_{i,1}$ given by

$$\theta_{i,0}(x_i) = 1, \quad \theta_{i,0}(x_{i+1}) = 0,$$
$$\theta_{i,1}(x_i) = 0, \quad \theta_{i,1}(x_{i+1}) = 1.$$

The polynomials $\{\theta_{i,0}, \theta_{i,1}\}$ are called *the local shape functions* associated to the above local degrees of freedom for $\mathbb{P}_1(K_i)$. Example 2.31 implies that these local shape functions form a basis for $\mathbb{P}_1(K_i)$.

One can derive global shape functions (that is, the hat functions of Figure 3.2) by using local shape functions through *the finite element assembly process*: For each $K_i = [x_i, x_{i+1}] \in T_h$, we assign the local shape functions $\theta_{i,0}$ and $\theta_{i,1}$ to x_i and x_{i+1}, respectively. The global function ψ_i, which is nonzero at the vertex x_i of T_h, is obtained by noticing that ψ_i is nonzero only on those elements that contain x_i, that is, K_{i-1} and K_i. Then, the restriction of ψ_i to any of these elements is equal to the local shape function assigned to x_i. This means that

$$\psi_i|_{K_{i-1}} = \theta_{i-1,1}, \text{ and } \psi_i|_{K_i} = \theta_{i,0}.$$

For example, the global shape function ψ_2 in Figure 3.2 is nonzero on K_1 and K_2 with $\psi_2|_{K_1} = \theta_{1,1}$, and $\psi_2|_{K_2} = \theta_{2,0}$.

The above discussions imply that the minimum information that we need to locally construct global interpolations over a mesh T_h is the triplet $(K_i, \mathbb{P}_1(K_i), \Sigma_i)$ for any $K_i \in T_h$, where Σ_i contains local degrees of freedom. This triplet is called a *1D Lagrange finite element of type (1)*.

The fact that global results can be obtained by using local information is implemented in the FEniCS code of the previous section by the following segment:

```
CGE = FiniteElement("Lagrange", mesh.ufl_cell(), 1)
Z = FunctionSpace(mesh, CGE)
```

Lagrange elements are denoted by Lagrange in FEniCS. The first line of the above segment defines CGE as a Lagrange finite element of type (1), where `mesh.ufl_cell()` refers to K_i, and `"Lagrange"` and 1 specify the degrees of freedom and the local approximation space. The second line defines the global approximation space by using the mesh and the finite element CGE.

3.2 FINITE ELEMENTS

Motivated by the discussions of the previous section, *a finite element* is defined to be a triplet $(K, \mathbb{P}, \Sigma)$, where:

- The set K, also called element, is a bounded subset of $\mathbb{R}^d$. There is no limitation on the shape of K. For 1D problems, K is a closed interval. In 2D, K is usually a triangle or a rectangle. In 3D, K is usually a tetrahedron, a prism, or a (rectangular) cuboid.
- The set $\mathbb{P}$ is the local approximation space and is a finite-dimensional space of polynomials on K that can be $\mathbb{P}_k(K)$ or one of its subspaces.
- The set Σ is a set containing some information that can uniquely specify any member of $\mathbb{P}$. Mathematically speaking, $\Sigma = \{\sigma_1, \ldots, \sigma_N\}$ is a basis for the dual space $\mathbb{P}'$. The members σ_i, $i = 1, \ldots, N$, are called *the local degrees of freedom of the finite element* $(K, \mathbb{P}, \Sigma)$. Notice that the number of degrees of freedom N is the same as $\dim \mathbb{P}$. *The (local) shape functions or the basis functions of the finite element* $(K, \mathbb{P}, \Sigma)$ are the members of the dual basis of Σ. Therefore, local shape functions $\{\theta_1, \ldots, \theta_N\}$ are polynomials belonging to $\mathbb{P}$ that satisfy

$$\sigma_i(\theta_j) = \delta_{ij}, \ i, j = 1, \ldots, N. \tag{3.7}$$

Given a function f defined on K, *the local interpolation operator I_K* of the finite element $(K, \mathbb{P}, \Sigma)$ is defined as

$$I_K f = \sum_{i=1}^{N} \sigma_i(f) \theta_i,$$

where $I_K f$ is also called *the $\mathbb{P}$-interpolant* of the function f. Since $I_K f$ is a linear combination of the local shape functions, it belongs to $\mathbb{P}$. Notice that

$$\sigma_j(I_K f) = \sigma_j \left(\sum_{i=1}^{N} \sigma_i(f) \theta_i \right) = \sum_{i=1}^{N} \sigma_i(f) \sigma_j(\theta_i) = \sum_{i=1}^{N} \sigma_i(f) \delta_{ij} = \sigma_j(f),$$

that is, the values of degrees of freedom at f and at its $\mathbb{P}$-interpolant are the same. It is straightforward to show that $I_K p = p$, if $p \in \mathbb{P}$. We say two finite elements $(K, \mathbb{P}, \Sigma)$ and $(\tilde{K}, \tilde{\mathbb{P}}, \tilde{\Sigma})$ are equal if $K = \tilde{K}$, $\mathbb{P} = \tilde{\mathbb{P}}$, and $I_K = I_{\tilde{K}}$.

In the remainder of this section, we introduce a few examples of finite elements.

3.2.1 SIMPLICIAL LAGRANGE FINITE ELEMENTS OF TYPE (k)

Given a finite element $(K, \mathbb{P}, \Sigma)$ with $\dim \mathbb{P} = N$, suppose there are some points $b_1, \ldots, b_N \in K$ such that $\sigma_i(f) = f(b_i)$, that is, the degrees of freedom are the values of functions at the points $\{b_1,\ldots,b_N\}$. In this case, the points $\{b_1,\ldots,b_N\}$ are called the *nodes* of the finite element, $(K,\mathbb{P},\Sigma)$ is called *a Lagrange finite element*, and the local shape functions $\{\theta_1,\ldots,\theta_N\}$ are also called *the nodal basis of* $\mathbb{P}$. The associated local interpolation operator is given by $I_K f = \sum_{i=1}^{N} f(b_i)\theta_i$, that is, Lagrange interpolants are constructed by matching point values at the nodes of Lagrange finite elements.

In this section, a class of Lagrange elements on *simplices* is introduced. *An n-simplex* is a subset of $\mathbb{R}^n$ defined as follows: A 1-simplex is a closed interval $[a_0,a_1]$ with *vertices* $\{a_0,a_1\}$. A 2-simplex is a triangle with the vertices $\{a_0,a_1,a_2\}$. The *face* (in 2D also called *edge*) opposite to the vertex a_i is denoted by F_i. The *outward unit normal vector of* F_i is denoted by $\mathbf{n}_i$. A 3-simplex is a tetrahedron with the vertices $\{a_0,\ldots,a_3\}$. The face F_i and the outward unit normal vector $\mathbf{n}_i$ are defined similar to 2-simplices. Figure 3.6 shows an *n*-simplex for $n=1,2,3$.

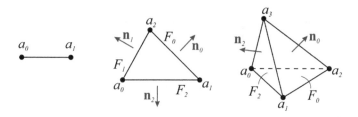

Figure 3.6   An *n*-simplex with the vertices $\{a_0,\ldots,a_n\}$ for $n=1$ (left), $n=2$ (center), and $n=3$ (right). The face opposite to the vertex a_i is denoted by F_i and the vector $\mathbf{n}_i$ is the outward unit normal vector of F_i.

A useful way for specifying a point x of an *n*-simplex is through its *barycentric coordinates* $\{\lambda_0,\ldots,\lambda_n\}$ given by

$$\lambda_i(\mathbf{x}) = 1 - \frac{(\mathbf{x}-a_i)\cdot\mathbf{n}_i}{(a_j-a_i)\cdot\mathbf{n}_i}, \tag{3.8}$$

where $\mathbf{x}=(x_1,\ldots,x_n)$, and a_j is any vertex in F_i. Notice that λ_i is the normalized distance of $\mathbf{x}$ from the face F_i: We have $\lambda_i = 1$ at a_i and $\lambda_i = 0$ at all points of F_i, see Figure 3.7. It is not hard to show that $0 \le \lambda_i(\mathbf{x}) \le 1$, and $\sum_{i=0}^{n}\lambda_i(\mathbf{x}) = 1$. The definition (3.8) implies that

$$\lambda_i(a_j) = \delta_{ij}, \; i,j=0,\ldots,n. \tag{3.9}$$

The point $\mathbf{x}$ of an *n*-simplex with $\lambda_i(\mathbf{x}) = \frac{1}{n+1}$, $i=0,\ldots,n$, is called *the barycenter* or *the center of gravity* of that simplex.

Example 3.1. *The unit 2-simplex is a triangle with the vertices* $a_0 = (0,0)$, $a_1 = (1,0)$, *and* $a_2 = (0,1)$, *see Figure 3.7. The barycentric coordinates of a point* $x = (x_1,x_2)$ *of the unit 2-simplex are given by*

$$\lambda_0(x_1,x_2) = 1-x_1-x_2, \; \lambda_1(x_1,x_2) = x_1, \text{ and } \lambda_2(x_1,x_2) = x_2.$$

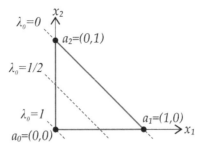

Figure 3.7 The barycentric coordinate λ_0 associated to the vertex $a_0 = (0,0)$ of the unit 2-simplex.

Note that, generally speaking, λ_i is an affine function of $x_1, \ldots, x_n$.

Example 3.2. *The unit* 3-*simplex* is a tetrahedron with the vertices $a_0 = (0,0,0)$, $a_1 = (1,0,0)$, and $a_2 = (0,1,0)$, and $a_3 = (0,0,1)$. The definition (3.8) implies that the barycentric coordinates of a point $x = (x_1, x_2, x_3)$ of the unit 3-simplex are given by

$$\lambda_0(x_1, x_2, x_3) = 1 - x_1 - x_2 - x_3, \text{ and } \lambda_i(x_1, x_2, x_3) = x_i, \ i = 1, 2, 3.$$

The simplest Lagrange finite element on a 1-simplex $K = [a_0, a_1]$ is the 1D Lagrange finite element of type (1), also called 1-*simplex of type* (1), that is introduced in Section 3.1.2.

To introduce the simplest Lagrange finite element on a 2-simplex K, we first need to show that any polynomial $p \in \mathbb{P}_1(K)$ can be uniquely determined by its values at the vertices $\{a_0, a_1, a_2\}$ of K. Notice that $\dim \mathbb{P}_1(K) = 3$, and that (3.8) implies that $\lambda_i \in \mathbb{P}_1(K)$, $i = 0, 1, 2$. Suppose $f(x) = \sum_{i=0}^{2} c_i \lambda_i(x) = 0$, for all $x \in K$. By using (3.9), one concludes that $f(a_j) = c_j$. Thus, $f(x) = 0$, suggests that $c_0 = c_1 = c_2 = 0$, and therefore, $\{\lambda_0, \lambda_1, \lambda_2\}$ is a basis for $\mathbb{P}_1(K)$. Let $p(x) = \sum_{i=0}^{2} c_i \lambda_i(x)$. Again (3.9) implies that $p(a_j) = c_j$. Consequently, any $p \in \mathbb{P}_1(K)$ can be uniquely determined by its values at the vertices of K with $p(x) = \sum_{i=0}^{2} p(a_i) \lambda_i(x)$.

The simplest Lagrange finite element on a 2-simplex K with the vertices $\{a_0, a_1, a_2\}$ is 2-*simplex (or triangle) of type (1)* $(K, \mathbb{P}_1(K), \Sigma_K)$, where the degrees of freedom $\Sigma_K = \{\sigma_0, \sigma_1, \sigma_2\}$ are values at the vertices of K given by $\sigma_i(p) = p(a_i)$. By abusing the notation, we also write $\Sigma_K = \{p(a_0), p(a_1), p(a_2)\}$. The relations (3.7) and (3.9) imply that the local shape functions associated to Σ_K are $\{\lambda_0, \lambda_1, \lambda_2\}$. The local interpolation operator for n-simplex of type (1) with $n = 2$ is given by

$$I_K^1 f = \sum_{i=0}^{n} f(a_i) \lambda_i. \tag{3.10}$$

Conventional finite element diagrams are usually employed to depict finite elements. These diagrams show the shape of an element and its degrees of freedom. Figure 3.8 depicts the conventional finite element diagram of 2-simplex of type (1).

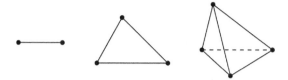

Figure 3.8 The conventional finite element diagrams of n-simplex of type (1) with $n = 1$ (left), $n = 2$ (center), and $n = 3$ (right).

Similarly, we can define 3-*simplex (or tetrahedron) of type (1)* $(K, \mathbb{P}_1(K), \Sigma_K)$, where K is a tetrahedron with vertices $\{a_0, \ldots, a_3\}$ and $\Sigma_K = \{p(a_0), \ldots, p(a_3)\}$ are values at the vertices of K. The local shape functions are $\{\lambda_0, \ldots, \lambda_3\}$ and the associated local interpolation operator is given by (3.10) with $n = 3$. The conventional finite element diagram of tetrahedron of type (1) is shown in Figure 3.8.

More generally, given an n-simplex K with $n = 1, 2, 3$, and any integer $k \geq 1$, one can define a Lagrange finite element $(K, \mathbb{P}_k(K), \Sigma_K)$ called n-*simplex of type* (k), where $\mathbb{P}_k(K)$ is the space of polynomials of degree k on K and $\Sigma_K = \{p(b_0), \ldots, p(b_{N-1})\}$, with $N = \dim \mathbb{P}_k(\mathbb{R}^n) = \binom{n+k}{k}$. The nodes $\{b_0, \ldots, b_{N-1}\}$ have the barycentric coordinates

$$\left(\frac{j_0}{k}, \ldots, \frac{j_n}{k}\right), \tag{3.11}$$

where $j_0, \ldots, j_n$ are any choice of non-negative integers such that $j_0 + \cdots + j_n = k$, and $j_i < k + 1$, $i = 0, \ldots, n$. The local interpolation operator associated to n-simplex of type (k) is denoted by I_K^k.

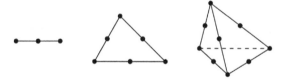

Figure 3.9 The conventional finite element diagrams of n-simplex of type (2) with $n = 1$ (left), $n = 2$ (center), and $n = 3$ (right). Only visible degrees of freedom are shown.

Example 3.3. Let $n = 1$ and $k = 2$. Then, the barycentric coordinates of the nodes of 1-simplex of type (2) are given by $b_0 = (1, 0)$, $b_1 = (0, 1)$, and $b_2 = (\frac{1}{2}, \frac{1}{2})$. The conventional finite element diagram of this element is shown in Figure 3.9. The shape functions can be written as $\theta_0(x) = \lambda_0(x)(1 - 2\lambda_1(x))$, $\theta_1(x) = \lambda_1(x)(1 - 2\lambda_0(x))$, and $\theta_2(x) = 4\lambda_0(x)\lambda_1(x)$. These shape functions are shown in Figure 3.10.

Example 3.4. The barycentric coordinates of the nodes of 2-simplex of type (2) are

$$b_0 = (1, 0, 0), \qquad b_1 = (1/2, 1/2, 0), \quad b_2 = (0, 1, 0),$$
$$b_3 = (1/2, 0, 1/2), \quad b_4 = (0, 1/2, 1/2), \quad b_5 = (0, 0, 1).$$

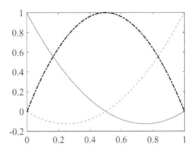

Figure 3.10 Local shape functions of 1-simplex of type (2) on $K = [0,1]$.

Figure 3.9 shows the conventional finite element diagram of this finite element. It is not hard to show that the associated local shape functions are given by

$$\theta_0 = \lambda_0(2\lambda_0 - 1), \quad \theta_1 = 4\lambda_0\lambda_1, \quad \theta_2 = \lambda_1(2\lambda_1 - 1),$$
$$\theta_3 = 4\lambda_0\lambda_2, \quad\quad\quad \theta_4 = 4\lambda_1\lambda_2, \quad \theta_5 = \lambda_2(2\lambda_2 - 1).$$

Simplicial Lagrange finite elements are available in FEniCS. For examples, the following code defines n-simplex of type (2) for $n = 1, 2, 3$.

```
FE_1simplex = FiniteElement("Lagrange", "interval", 2)
FE_2simplex = FiniteElement("Lagrange", "triangle", 2)
FE_3simplex = FiniteElement("Lagrange", "tetrahedron", 2)
```

To define simplicial Lagrange elements in FEniCS, one can also use "CG" instead of "Lagrange" in the above code.

3.2.2 SIMPLICIAL HERMITE FINITE ELEMENTS OF TYPE (3)

Suppose degrees of freedom of a finite element $(K, \mathbb{P}, \Sigma)$ are values of functions at some points $\{b_1, \ldots, b_v\} \subset K$ together with values of partial derivatives (or directional derivatives) of functions at some points $\{\bar{b}_1, \ldots, \bar{b}_d\} \subset K$. The finite element $(K, \mathbb{P}, \Sigma)$ is then called *a Hermit finite element* and the points $\{b_1, \ldots, b_v, \bar{b}_1, \ldots, \bar{b}_d\}$ are called the nodes of the finite element. In the following, we define simplicial Hermite finite elements that involve polynomials of degree 3. Since we will not use 2D and 3D Hermite elements in the remainder of this book, we only discuss the 1D case in some details.

Let $K = [a_0, a_1]$ be a 1-simplex. *The Hermite 1-simplex of type (3)* is given by $(K, \mathbb{P}_3(K), \Sigma_K)$ with $\Sigma_K = \{p(a_0), p'(a_0), p(a_1), p'(a_1)\}$, where $p'(x)$ is the derivative of $p(x)$ with respect to x. The conventional finite element diagram of this element is shown in Figure 3.11, where circles centered at vertices indicate degrees of freedom associated to derivatives at those vertices. It is not hard to show that the local

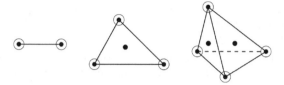

Figure 3.11 The conventional finite element diagrams of the Hermite n-simplex of type (3) with $n = 1$ (left), $n = 2$ (center), and $n = 3$ (right). Only visible degrees of freedom are shown.

shape functions can be written as

$$\theta_0(x) = \frac{1}{h^3}(x - a_1)^2(h + 2(x - a_0)), \quad \theta_1(x) = \frac{1}{h^2}(x - a_0)(x - a_1)^2,$$

$$\theta_2(x) = \frac{1}{h^3}(x - a_0)^2(h - 2(x - a_1)), \quad \theta_3(x) = \frac{1}{h^2}(x - a_1)(x - a_0)^2, \tag{3.12}$$

where $h = a_1 - a_0$. Figure 3.12 shows these local shape functions on the unit interval.

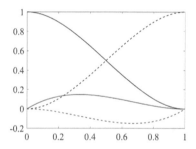

Figure 3.12 Local shape functions of the Hermite 1-simplex of type (3) on $K = [0,1]$.

Suppose K is a 2-simplex with the vertices $\{a_0, a_1, a_2\}$. *The Hermite 2-simplex (or triangle) of type (3) is* $(K, \mathbb{P}_3(K), \Sigma_K)$, where Σ_K includes values of functions and all their first-order partial derivatives at the vertices together with the value of functions at the node $\frac{1}{3}(a_0 + a_1 + a_2)$. *The Hermite 3-simplex (or tetrahedron) of type (3)* can be defined in a similar way. The conventional finite element diagram of these elements are shown in Figure 3.11. Hermite elements are still not available in FEniCS.

3.2.3 THE RAVIART-THOMAS FINITE ELEMENT

Let K be a 2-simplex and suppose $\mathbb{RT}$ is the space of 2D polynomial vector fields of the form $\boldsymbol{p}(x_1, x_2) = (c_0 + c_2 x_1, c_1 + c_2 x_2)$, where c_0, c_1, and c_2 are constant. It is not hard to show that $\mathbb{RT}$ is a linear space with $\dim \mathbb{RT} = 3$. Referring to Figure 3.6, for any $\boldsymbol{p} \in \mathbb{RT}$, let $\sigma_i(\boldsymbol{p})$ be the "flux" of $\boldsymbol{p}$ across the face F_i, that is,

$$\sigma_i(\boldsymbol{p}) = \int_{F_i} \boldsymbol{p} \cdot \mathbf{n}_i \, ds. \tag{3.13}$$

One can show that the values of these 3 fluxes uniquely specify polynomials belonging to $\mathbb{RT}$. Therefore, $(K, \mathbb{RT}, \Sigma_K)$ with $\Sigma_K = \{\sigma_0, \sigma_1, \sigma_2\}$, is a finite element which is called *the (2D) Raviart-Thomas finite element*. The conventional finite element diagram of this element is shown in Figure 3.13. To motivate the definition of the local shape functions of this element, let us consider the following example.

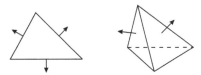

Figure 3.13 The conventional finite element diagram of the 2D (left) and the 3D (right) Raviart-Thomas finite elements. Arrows normal to faces denote the degrees of freedom associated to faces. Only visible degrees of freedom are shown.

Example 3.5. Let $x = (x_1, x_2)$, and consider the vector field $\boldsymbol{\theta}_0 \in \mathbb{RT}$ given by

$$\boldsymbol{\theta}_0(x_1, x_2) = \frac{1}{2A_K}(x - a_0),$$

where A_K and a_0 are respectively the area and a vertex of the 2-simplex K. Using the notation of Figure 3.6, any point x on the face F_0 can be denoted as $x = a_1 + t(a_2 - a_1)$, $0 \le t \le 1$. Thus, on F_0 we can write

$$\boldsymbol{\theta}_0(x) \cdot \mathbf{n}_0 = \frac{1}{2A_K}(a_1 + t(a_2 - a_1) - a_0) \cdot \mathbf{n}_0 = \frac{(a_1 - a_0) \cdot \mathbf{n}_0}{2A_K},$$

where we used the fact that $a_2 - a_1$ is normal to $\mathbf{n}_0$. Since $(a_1 - a_0) \cdot \mathbf{n}_0$ is equal to the height h_0 of K normal to F_0, we conclude that the component of $\boldsymbol{\theta}_0(x)$ normal to F_0 is constant and is given by $\boldsymbol{\theta}_0(x) \cdot \mathbf{n}_0 = 1/|F_0|$, where $|F_0|$ is the length of F_0. Consequently, one concludes that

$$\sigma_0(\boldsymbol{\theta}_0) = \int_{F_0} \boldsymbol{\theta}_0 \cdot \mathbf{n}_0 \, ds = \frac{1}{|F_0|} \int_{F_0} ds = 1.$$

Similarly, one can show that $\sigma_1(\boldsymbol{\theta}_0) = \sigma_2(\boldsymbol{\theta}_0) = 0$.

The above example suggests that the local shape functions of the 2D Raviart-Thomas finite element are given by

$$\boldsymbol{\theta}_i(x_1, x_2) = \frac{1}{2A_K}(x - a_i), \quad i = 0, 1, 2.$$

Figure 3.14 shows these local shape functions. The local interpolation operator of the 2D Raviart-Thomas finite element can be written as

$$I_K^{\mathrm{RT}} v = \sum_{i=0}^{2} \left(\int_{F_i} v \cdot \mathbf{n}_i \, ds \right) \boldsymbol{\theta}_i.$$

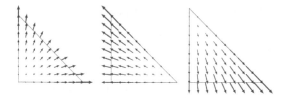

Figure 3.14 The local shape functions of the 2D Raviart-Thomas finite element.

The 3D Raviart-Thomas finite element is obtained by using local degrees of freedom similar to (3.13). Figure 3.13 shows the conventional finite element diagram of this 3D finite element. More generally, it is also possible to define 2D and 3D Raviart-Thomas finite elements of degree k, for any integer $k \geq 1$. In FEniCS, the 2D and 3D Raviart-Thomas finite elements (of degree 1) can be defined as follows:

```
RT_2D = FiniteElement("RT", "triangle", 1)
RT_3D = FiniteElement("RT", "tetrahedron", 1)
```

3.2.4 THE NEDELEC FINITE ELEMENT

Suppose K is a 2-simplex and let $\mathbf{t}_i$ be one of the two unit vectors parallel to the edge E_i of K (in 2D, edges and faces are the same). Let us denote the space of all polynomial vector fields of the form $\mathbf{p}(x_1, x_2) = (c_0 + c_2 x_2, c_1 - c_2 x_1)$, with c_0, c_1, and c_2 being constant, by $\mathbb{NE}$. One can show that $\mathbb{NE}$ is a linear space with $\dim \mathbb{NE} = 3$. For any $\mathbf{p} \in \mathbb{NE}$, suppose $\sigma_i(\mathbf{p})$ is the integral of the tangent component of $\mathbf{p}$ along E_i, that is,

$$\sigma_i(\mathbf{p}) = \int_{E_i} \mathbf{p} \cdot \mathbf{t}_i \, ds. \tag{3.14}$$

The 2D Nédélec finite element (of the first kind) $(K, \mathbb{NE}, \Sigma_K)$ is obtained by assuming $\Sigma_K = \{\sigma_0, \sigma_1, \sigma_2\}$. The conventional finite element diagram of this element is shown in Figure 3.15.

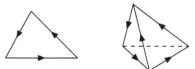

Figure 3.15 The conventional finite element diagram of the 2D (left) and the 3D (right) Nédélec finite elements. Arrows parallel to edges denote the degrees of freedom associated to edges. Only visible degrees of freedom are shown.

The local shape functions of this element can be defined as follows: Let $R(x_1, x_2) = (x_2, -x_1)$, that is, $R(x_1, x_2)$ is obtained by a 90 degree rotation of $\mathbf{x} = (x_1, x_2)$ in the clockwise direction. Then, by using an approach similar to that of

Example 3.5, one can show that the local shape functions are given by

$$\boldsymbol{\theta}_i(x_1,x_2) = \frac{R(\boldsymbol{x}-a_l)}{|E_i|\mathbf{t}_i \cdot R\left(\frac{1}{2}(a_{i_1}+a_{i_2})-a_i\right)}, \quad i=0,1,2,$$

where i_1 and i_2 are not equal to i. Figure 3.16 depicts these local shape functions. The local interpolation operator of the 2D Nédélec finite element reads

$$I_K^N \boldsymbol{v} = \sum_{i=0}^{2}\left(\int_{E_i} \boldsymbol{v}\cdot\mathbf{t}_i\,ds\right)\boldsymbol{\theta}_i.$$

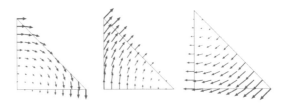

Figure 3.16   The local shape functions of the 2D Nédélec finite element.

Local degrees of freedom of the 3D Nédélec finite element is defined similar to (3.14). Figure 3.15 shows the conventional finite element diagram of this 3D finite element. More generally, one can define 2D and 3D Nédélec finite elements of degree k, for any integer $k \geq 1$. In FEniCS, the 2D and 3D Nédélec finite elements (of degree 1) can be defined by the following code:

```
ND_2D = FiniteElement("N1curl", "triangle", 1)
ND_3D = FiniteElement("N1curl", "tetrahedron", 1)
```

3.3   MESHES

To construct approximation spaces over a domain Ω by using finite elements, one needs to employ *a mesh* of Ω. More specifically, let Ω be an open, bounded subset of $\mathbb{R}^n$ with the continuous boundary $\partial\Omega$ and let $\overline{\Omega} = \Omega \cup \partial\Omega$. In this book, domains are usually assumed to be polygons for 2D problems and polyhedrons for 3D problems. A mesh of Ω is a set $\{K_1,\ldots,K_m\}$, where

1. Each K_i, called *an element* or *a cell* of the mesh, is usually assumed to be a polygonal domain in 2D or a polyhedral domain in 3D that contains its boundary;
2. $\overline{\Omega} = K_1 \cup K_2 \cup \cdots \cup K_m$;
3. Two different elements either do not intersect or their intersection is a common vertex in 1D, a common vertex or edge in 2D, or a common vertex, edge, or face in 3D. Later, we will see this property of a mesh is important for obtaining approximation spaces over Ω.

Meshes defined as above are also called *geometrically conformal meshes*. We consider meshes with simplicial elements in this book, that is, elements are triangles in 2D and tetrahedrons in 3D. These meshes are called *triangulations* as well. For example, the left panel of Figure 3.17 shows a mesh of a square. The configuration on the right panel of Figure 3.17 is not a mesh since the intersection of each of the two smaller triangles with the bigger one is not an edge of the bigger triangle. We denote the number of vertices, edges, faces, and elements of a mesh by n_v, n_e, n_f, and n_{el}, respectively. Notice that in 2D, edges and faces are the same.

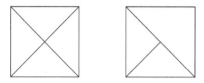

Figure 3.17   An example (left) and a counterexample (right) of a mesh.

Given a mesh $\{K_1, \ldots, K_m\}$, let $h_{K_i} = \operatorname{diam} K_i$, be *the diameter of the element* K_i, that is, h_{K_i} is the maximum distance of two different points of K_i. The mesh is usually denoted by T_h, where $h = \max\{h_{K_1}, \ldots, h_{K_m}\}$. The parameter h refers to the refinement level of the mesh. To numerically approximate solutions, one usually considers a sequence $T_{h_1}, T_{h_2}, T_{h_3}, \ldots$ of successively refined meshes of a domain with $h_1 > h_2 > h_3 > \cdots$. This sequence is denoted by $\{T_h\}_{h>0}$ or simply $\{T_h\}$, see Figure 3.18.

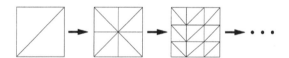

Figure 3.18   A sequence of meshes of a square.

Meshes may be generated by using *a reference element*. For example, consider the mesh $T_h = \{K_1, K_2, K_3\}$ shown in Figure 3.19 and let $\hat{K}$ be an arbitrary triangle. There are affine transformations $f_i : \hat{K} \to K_i$, $i = 1, 2, 3$, that map $\hat{K}$ onto K_i with $f_i(\mathbf{x}) = \mathbf{v}_i + M_i\mathbf{x}$, where $\mathbf{v}_i$ is a vector in $\mathbb{R}^2$ and M_i is a 2×2 matrix. Therefore, the affine transformations $\{f_1, f_2, f_3\}$ allow to generate the mesh T_h from the reference element $\hat{K}$. Such meshes are also called *affine meshes*.

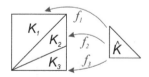

Figure 3.19   An affine mesh can be generated from a reference element $\hat{K}$.

Notice that if Ω has a curved boundary, it is not possible to completely cover it by meshes that only contain elements with flat faces. In this case, one can use meshes with flat-face elements to approximately cover Ω. One can improve this approximation by using finer meshes at curved boundaries. Another approach is to use meshes with curved-face elements. We will not consider curved-face elements in this book.

Meshes of some simple geometries can be directly defined in FEniCS. For example, the following code generates structured meshes of the unit square and the unit cube:

```
Mesh_S = UnitSquareMesh(4,4)
Mesh_C = UnitCubeMesh(4,4,4)
```

The numbers inside parenthesis show the number of divisions of each edge. Figure 3.20 shows these meshes.

Figure 3.20   Structured meshes of the unit square and the unit cube.

Unstructured meshes and meshes for more complex geometries can be defined by the mesh generation component of FEniCS called `mshr`. The following segment generates the unstructured meshes of Figure 3.21.

```
from mshr import *

# a rectangle with a rectangular hole
Rec1 = Rectangle(Point(0, 0), Point(1, 1))
Rec2 = Rectangle(Point(0.25, 0.25), Point(0.75, .75))
mesh_Rec = generate_mesh(Rec1 - Rec2, 8)

# a triangle
vertices = [Point(0.0, 0.0), Point(1.0, 0.0), Point(0.5, 1.0)]
Tri = Polygon(vertices)
mesh_Tri = generate_mesh(Tri, 6)

# the unit cube
Cube = Box(Point(0, 0, 0), Point(1, 1, 1))
mesh_Cube = generate_mesh(Cube, 7.5)

# the unit sphere centered at the origin
Spr = Sphere(Point(0, 0, 0), 1.0)
mesh_Spr = generate_mesh(Spr, 6)
```

Notice that the second argument of the function `generate_mesh` determines the refinement level of unstructured meshes and does not need to be an integer. Another option for defining meshes in FEniCS is to import meshes that are generated by other programs outside FEniCS.

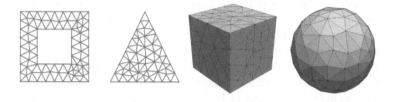

Figure 3.21   Unstructured meshes of some 2D and 3D domains.

It is easy to recover basic properties of meshes in FEniCS. The following code prints some properties of the mesh of the unit square shown in Figure 3.20.

```
n_el = Mesh_S.num_cells()    # number of elements
n_f = Mesh_S.num_faces()     # number of faces
n_v = Mesh_S.num_vertices() # number of vertices
h = Mesh_S.hmax() # the maximum diameter of elements

print("n_el = %.f, n_f = %.f, n_v = %.f, h = %.3f"
      % (n_el, n_f, n_v, h))
```

The output of this code is:

```
n_el = 32, n_f = 32, n_v = 25, h = 0.354
```

3.4   FINITE ELEMENT SPACES AND INTERPOLATIONS

In this section, we use finite elements to construct approximation spaces. Such approximation spaces are called *finite element spaces*. We begin this section by discussing this construction in its full generality suitable for any finite element. We then apply this construction to finite elements introduced earlier. The reader is encouraged to read this general discussion again while studying the construction for each specific element.

Let $T_h = \{K_1, \ldots, K_m\}$ be a mesh of a domain Ω and let $\{(K, \mathbb{P}_K, \Sigma_K)\}_{K \in T_h}$ be a family of finite elements associated to the elements of T_h. The finite element space induced by this family of finite elements is constructed through *the finite element assembly process* as follows:

1. Each local degree of freedom of each finite element $(K, \mathbb{P}_K, \Sigma_K)$ is assigned to a geometric entity of the mesh T_h such as a vertex, a node, an edge, or a face of T_h. Notice that several local degrees of freedom may be assigned to a geometric entity.

2. The finite element space V_h associated to T_h is defined as the linear space of functions f such that $f|_K \in \mathbb{P}_K$, for all $K \in T_h$, and $\sigma_{K,i}(f|_K) = \sigma_{K',j}(f|_{K'})$, if the i-th local degree of freedom $\sigma_{K,i}$ of K and the j-th local degree of freedom $\sigma_{K',j}$ of K' are assigned to the same geometric entity of T_h.

The members of any basis for the linear space V_h are called *global shape functions*. An important property of the finite element space V_h is that it admits *locally supported* global shape functions, that is, bases functions that are zero everywhere on T_h except for a few elements. *These global shape functions can be constructed by using local shape functions of finite elements.* Before discussing this construction, let us show that finite element spaces are finite-dimensional. Let n_t be the total number of the geometric entities of T_h that were assigned to at least one local degree of freedom in the finite element assembly process. For any $v \in V_h$, all local degrees of freedom assigned to the same geometric entity of T_h take the same value at v. Consequently, any $v \in V_h$ can be uniquely specified by using $\{g_1, \ldots, g_{n_t}\}$, where $g_i(v)$ assigned to the i-th geometric entity is the common value of all local degrees of freedom assigned to the i-th geometric entity at v. *This argument suggests that the linear space V_h is a finite-dimensional space with* $\dim V_h = n_t$.

One can obtain a set of locally supported global shape functions $\{\psi_1, \ldots, \psi_{n_t}\}$, which is the dual basis of $\{g_1, \ldots, g_{n_t}\}$ in the sense that $g_i(\psi_j) = \delta_{ij}$, as follows: Suppose that only 2 local degrees of freedom $\sigma_{K,j}$ and $\sigma_{K',l}$ are associated to the i-th geometric entity of T_h. Let the local shape functions $\theta_{K,j}$ on K and $\theta_{K',l}$ on K' satisfy $\sigma_{K,j}(\theta_{K,j}) = \sigma_{K',l}(\theta_{K',l}) = 1$. Then, ψ_i is equal to: $\theta_{K,j}$ on K; $\theta_{K',l}$ on K'; and zero on other elements of T_h. The same approach also holds if the i-th geometric entity is shared by more than 2-elements. It is straightforward to show the duality relation $g_i(\psi_j) = \delta_{ij}$. This relation implies that $\{\psi_1, \ldots, \psi_{n_t}\}$ is a basis for V_h. *The (global) interpolation $I_h f \in V_h$ of a function f on Ω is then defined as*

$$I_h f = \sum_{i=1}^{n_t} g_i(f)\psi_i. \tag{3.15}$$

It is not hard to show that on any element $K \in T_h$, the global interpolant $I_h f$ satisfies

$$(I_h f)|_K = I_K(f|_K),$$

where I_K is the local interpolation operator of the finite element $(K, \mathbb{P}_K, \Sigma_K)$.

If the finite element space V_h is a subset of a linear space V, we say V_h is a *V-conformal space*. In the remainder of this section, we will employ the finite elements introduced in Section 3.2 to derive H^1-, $H(\text{div})$-, and $H(\text{curl})$-conformal finite element spaces.

3.4.1 H^1-CONFORMAL FINITE ELEMENT SPACES

Recall that the Sobolev space $H^1(\Omega)$ introduced in Example 2.8 is the space of square integrable functions that also admit square integrable first-order derivatives. Given a mesh T_h of Ω, suppose a function $f : \Omega \to \mathbb{R}$ has the property that $f|_K \in C^1(K)$, for all $K \in T_h$. *One can show that $f \in H^1(\Omega)$ if and only if $f \in C^0(\overline{\Omega})$*. In the remainder of this section, we obtain finite element spaces associated to the simplicial Lagrange finite elements of type (k) and the simplicial Hermite finite element of type (3). We use the above result to show that these finite element spaces are H^1-conformal.

3.4.1.1 Lagrange Elements

Let T_h be a mesh of Ω with simplicial elements and suppose $\{(K, \mathbb{P}_k(K), \Sigma_K)\}_{K \in T_h}$ is a family of simplicial Lagrange finite elements. To obtain the associated finite element space $\mathbb{P}_k(T_h)$ through the finite element assembly process, we first assign local degrees of freedom of each K to the corresponding nodes of T_h. Then, $\mathbb{P}_k(T_h)$ is defined to be the space of functions f such that $f|_K \in \mathbb{P}_k(K)$, for all $K \in T_h$, and $f|_K(b_i) = f|_{K'}(b_i)$, whenever b_i is a common node of K and K'.

Let F be an $(n-1)$-dimensional face of an n-simplex K. We have $\dim \mathbb{P}_k(K) = \binom{n+k}{k}$, and $\dim \mathbb{P}_k(F) = \binom{n+k-1}{k}$. Notice that if $p \in \mathbb{P}_k(K)$, then $p|_F \in \mathbb{P}_k(F)$. It is not hard to show that n-simplex of type (k) has $\binom{n+k-1}{k}$ nodes on each face of K. Now, suppose F is a common face of K and K', and let $f \in \mathbb{P}_k(T_h)$. Then, *the requirement that $f|_K$ and $f|_{K'}$ have the same values at all common nodes on F implies that $f|_K = f|_{K'}$ on F.* Hence, we conclude that

$$\mathbb{P}_k(T_h) = \{f : f|_K \in \mathbb{P}_k(K), \text{ for all } K \in T_h\} \cap C^0(\overline{\Omega}).$$

The result mentioned at the beginning of this section then suggests that $\mathbb{P}_k(T_h)$ is an H^1-conformal finite element space. Let us consider a few examples.

Example 3.6. Let $T_h = \{K_0, \ldots, K_{N-1}\}$ be a mesh of an open interval Ω with the vertices $\{x_0, \ldots, x_N\}$ as discussed in Section 3.1.1. Also let $\{(K, \mathbb{P}_1(K), \Sigma_K)\}_{K \in T_h}$ be a family of 1-simplices of type (1). The conventional finite element diagram shown in Figure 3.8 tells us that local degrees of freedom are associated to the two vertices of each K. Therefore, two local degrees of freedom are assigned to the internal vertices $\{x_1, \ldots, x_{N-1}\}$ and only one local degrees of freedom is assigned to x_0 and x_N. Then, the associated finite element space $\mathbb{P}_1(T_h)$ is the space of continuous, piecewise affine functions over T_h defined in (3.1), with $\dim \mathbb{P}_1(T_h) = N+1$, as local degrees of freedom are assigned to $N+1$ vertices of T_h. Locally supported global shape functions $\{\psi_0, \ldots, \psi_N\}$ of $\mathbb{P}_1(T_h)$ are as follows: The basis function ψ_i associated to the vertex x_i is nonzero only at K_{i-1} and K_i, where it is equal to the local shape functions assigned to x_i, see Figure 3.2. Notice that in Section 3.1, the global shape functions for $\mathbb{P}_1(T_h)$ were obtained similarly.

Example 3.7. Let T_h be a mesh of a 2D domain and suppose $\{(K, \mathbb{P}_2(K), \Sigma_K)\}_{K \in T_h}$ is a family of 2-simplices of type (2). To obtain the associated finite element space $\mathbb{P}_2(T_h)$, the conventional finite element diagram of 2-simplex of type (2) shown in Figure 3.9 tells us that local degrees of freedom should be assigned to the vertices of T_h and the nodes at the middle of the edges of T_h, see Figure 3.22. The dimension

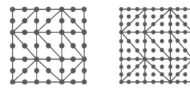

Figure 3.22   Nodes of a 2D mesh that are used in the finite element assembly process associated to 2-simplices of type (*k*) with $k = 2$ (left) and $k = 3$ (right).

of $\mathbb{P}_2(T_h)$ is the number of nodes of T_h that are assigned a local degree of freedom, that is, $\dim \mathbb{P}_2(T_h) = n_v + n_e$, where n_v and n_e are respectively the number of vertices and the number of edges of T_h. Locally supported global shape functions $\{\psi_1, \ldots, \psi_{n_v+n_e}\}$ are obtained by using local shape functions. Figure 3.23 shows regions where global shape functions associated to some nodes of the mesh of Figure 3.22 are not zero.

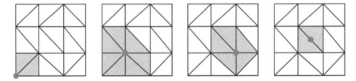

Figure 3.23   Supports of global shape functions induced by 2-simplices of type (2): The global shape function associated to the specified node is not zero on the shaded region.

Example 3.8. Consider a family of 2-simplices of type (3) on a 2D mesh. The dimension of the associated finite element space is $n_v + 2n_e + n_{el}$, where n_{el} is the number of elements of the mesh, see Figure 3.22.

Notice that the interelement continuity of functions belonging to finite element spaces is closely related to the third property of meshes discussed in Section 3.3. For example, the finite element assembly process does not result in continuous functions if one considers a family of 2-simplices of type (1) on the configuration on the right panel of Figure 3.17, see also Figure 3.24.

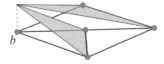

Figure 3.24   The global shape function for a family of 2-simplices of type (1) associated to the vertex *b* of the counterexample on the right panel of Figure 3.17 is not continuous.

By using locally supported global shape functions $\{\psi_1, \ldots, \psi_{n_t}\}$ and the general relation (3.15), one can write *the (global) Lagrange interpolation operator* I_h^k

induced by n-simplices of type (k) as

$$I_h^k f = \sum_{i=1}^{n_t} f(b_i)\psi_i,$$

where $b_1, \ldots, b_{n_t}$ are the nodes of the underlying mesh that have been used in the assembly process.

In Section 3.1.1, we saw how to employ FEniCS to interpolate functions using 1D Lagrange elements. Interpolations in higher dimensions are very similar. For example, the following code computes Lagrange interpolants of type (1) for $f(x, y) = 1 + 2^x y^2$ by using structured and unstructured meshes of the unit square.

```
# defining the function
f = Expression('1.0 + pow(2,x[0])*pow(x[1],2)', degree = 5)

# defining meshes
mesh_St = UnitSquareMesh(5,5) # structured mesh

Rec = Rectangle(Point(0, 0), Point(1, 1))
mesh_Unst = generate_mesh(Rec, 5) # unstructured mesh

# defining approximation spaces
CGE  = FiniteElement("Lagrange", "triangle", 1)
Z_St = FunctionSpace(mesh_St, CGE)
Z_Unst = FunctionSpace(mesh_Unst, CGE)

# interpolating
Interpolant_St = interpolate(f, Z_St)      # on structured mesh
Interpolant_Unst = interpolate(f, Z_Unst) # on unstructured mesh

# saving results in VTK format
vtkfile = File('Data/CGInterpolation_2DSt.pvd')
vtkfile << Interpolant_St

vtkfile = File('Data/CGInterpolation_2DUnSt.pvd')
vtkfile << Interpolant_Unst
```

This code works similar to that of Section 3.1.1. To define the approximation spaces Z_St and Z_Unst, we can also use the following alternative approach which is more compact:

```
Z_St = FunctionSpace(mesh_St, "Lagrange", 1)
Z_Unst = FunctionSpace(mesh_Unst, "Lagrange", 1)
```

At the end, the interpolants are saved in the VTK format, which is suitable for the visualization of results. The outputs are plotted in Figure 3.25 by using ParaView[1]. The interpolants are polynomials of degree 1 on each element.

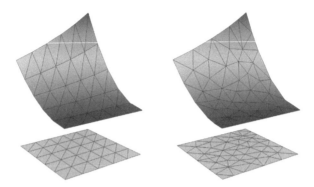

Figure 3.25   Lagrange interpolants of type (1) for $f(x,y) = 1 + 2^x y^2$ associated to structured (left) and unstructured (right) meshes of the unit square.

To examine the interpolants, we can obtain their values at different points and compare them with the values of the function $f(x,y)$ at those points. For example, the following code computes the values of the function and its interpolants at the point $(0.2, 0.4)$:

```
Point = (0.2, 0.4)
print("At point (%.1f,%.1f):" % (Point[0], Point[1]))
print("f(x,y) is:        %.10f" % f(Point))
print("Interpolant_St:   %.10f" % Interpolant_St(Point))
print("Interpolant_Unst: %.10f" % Interpolant_Unst(Point))
```

The output is:

```
At point (0.2,0.4):
f(x,y) is:        1.1837917368
Interpolant_St:   1.1837917368
Interpolant_Unst: 1.1900465961
```

Notice that contrary to `Interpolant_Unst`, the value of `Interpolant_St` is exactly equal to f at $(0.2, 0.4)$. The reason is that $(0.2, 0.4)$ is a vertex of the structured mesh but it is not a vertex of the unstructured mesh.

The above example involves scalar functions. Interpolations of vector fields using Lagrange elements in FEniCS is also straightforward. The main difference is that the definition of a scalar finite element space should be replaced with the following definition of a vectorial finite element space:

```
Z_vec = VectorFunctionSpace(mesh_St, "Lagrange", 1)
```

---

[1]http://www.paraview.org

3.4.1.2 Hermite Elements

Let T_h be a mesh of an interval Ω and suppose $\{(K,\mathbb{P}_3(K),\Sigma_K)\}_{K\in T_h}$ is a family of Hermite 1-simplices of type (3) introduced in Section 3.2.2. It is then straightforward to show that the associated finite element space V_h is the space of functions f such that $f|_K$ is a degree 3 polynomial for all elements K with $f|_K(b_i) = f|_{K'}(b_i)$, and $f'|_K(b_i) = f'|_{K'}(b_i)$, whenever b_i is a common vertex of K and K'. Then, V_h is an H^1-conformal finite element space as its members are continuous functions. In fact, V_h is also an H^2-conformal finite element space since the derivative of its members is also continuous.

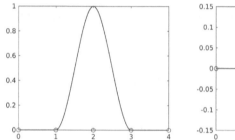 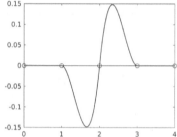

Figure 3.26 Global shape functions associated to the value (left) and the derivative (right) at the vertex 2 induced by a class of Hermite 1-simplices of type (3).

There are 2 degrees of freedom associated to each vertex of a Hermite 1-simplex of type (3) and therefore, $\dim V_h = 2n_v$. Global shape functions are obtained from the local shape functions depicted in Figure 3.12. Therefore, as shown in Figure 3.26, there are 2 global shape functions for each vertex of T_h. Notice that global shape functions and their derivatives are continuous. One can also derive finite element spaces for 2- and 3-simplices of type (3). The main difference is that in 2D and 3D, functions in the associated finite element spaces are continuous with discontinuous derivatives. Therefore, these finite element spaces are H^1-conformal but not H^2-conformal.

3.4.2 $H(\mathrm{div})$-CONFORMAL FINITE ELEMENT SPACES

Let $\{(K,\mathbb{RT},\Sigma_K)\}_{K\in T_h}$ be a family of 2D Raviart-Thomas finite elements on a mesh T_h. One can show that the associated finite element space V_h consists of vector fields $\boldsymbol{v}:\Omega\to\mathbb{R}^2$ with $\boldsymbol{v}|_K\in\mathbb{RT}$, for all $K\in T_h$, that satisfy the following interelement continuity: If a face F is a common face of the elements K and K', the components of $\boldsymbol{v}|_K$ and $\boldsymbol{v}|_{K'}$ at F which are normal to F are equal, see Figure 3.27. Then, one concludes that $V_h\subset H(\mathrm{div};\Omega)$. *Notice that unlike H^1-vector fields that are continuous, vector fields in $H(\mathrm{div};\Omega)$ are not continuous, in general.* Only the component of vector fields normal to each face is single-valued.

The conventional finite element diagram of the Raviart-Thomas finite element implies that for the assembly process, local degrees of freedom are assigned to

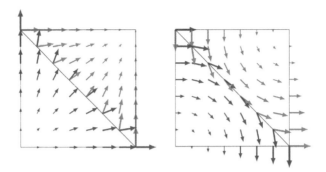

Figure 3.27 Global shape functions associated to the diagonal face induced by Raviart-Thomas (left) and Nédélec (right) elements.

all faces of T_h. Therefore, $\dim V_h = n_f$. Locally supported global shape functions $\{\boldsymbol{\psi}_1, \ldots, \boldsymbol{\psi}_{n_f}\}$ for V_h can be obtained by using local shape functions. To this end, the unit normal vector $\mathbf{n}_i$ is assigned to any face F_i, $i = 1, \ldots, n_f$ of T_h, where $\mathbf{n}_i$ is one of the two unit vectors normal to F_i. The global shape function associated to F_i is then obtained by using local shape functions compatible with $\mathbf{n}_i$. For example, see Figure 3.27. *The (global) Raviart-Thomas interpolation operator* is given by

$$ I_h^{\text{RT}} \boldsymbol{v} = \sum_{i=1}^{n_f} \left(\int_{F_i} \boldsymbol{v} \cdot \mathbf{n}_i \, ds \right) \boldsymbol{\psi}_i. $$

The above discussions also hold for the 3D Raviart-Thomas finite element.

In FEniCS, the interpolation of vector fields using the Raviart-Thomas element is very similar to the interpolation using Lagrange elements mentioned earlier. For example, the following code computes the Raviart-Thomas interpolant of degree 1 for the vector field $\boldsymbol{v}(x, y, z) = \left(y \sin z, \ x^2 \cos z, \ xy \right)$, over a mesh of the unit cube.

```
# defining the vector field
vf = Expression(("x[1]*sin(x[2])","pow(x[0],2)*cos(x[2])",
                "x[0]*x[1]"), degree = 5)

# defining mesh
Mesh = UnitCubeMesh(5,5,5)

# defining approximation space
Z_RT = FunctionSpace(Mesh, "RT", 1)

# interpolating
Interpolant_RT = interpolate(vf, Z_RT)
```

3.4.3 $H(\mathrm{curl})$-CONFORMAL FINITE ELEMENT SPACES

Let $\{(K, \mathbb{NE}, \Sigma_K)\}_{K \in T_h}$ be a family of 2D Nédélec finite elements on a mesh T_h. It is not hard to show that the finite element space V_h induced by this family is the space of vector fields $\boldsymbol{v} : \Omega \to \mathbb{R}^2$ with $\boldsymbol{v}|_K \in \mathbb{NE}$, for all $K \in T_h$, that satisfy the following interelement continuity: If an edge E is a common edge of the elements K and K', the components of $\boldsymbol{v}|_K$ and $\boldsymbol{v}|_{K'}$ at E which are parallel to E are equal, see Figure 3.27. One can show that $V_h \subset H(\mathrm{curl}; \Omega)$. *Thus, similar to $H(\mathrm{div})$-vector fields, vector fields in $H(\mathrm{curl}; \Omega)$ may not be continuous.* Only the tangent component of $H(\mathrm{curl})$-vector fields at each edge is single-valued.

The conventional finite element diagram of the Nédélec finite element suggests that for the assembly process, local degrees of freedom are assigned to all edges of T_h and therefore, $\dim V_h = n_\mathrm{e}$. To obtain locally supported global shape functions $\{\boldsymbol{\psi}_1, \ldots, \boldsymbol{\psi}_{n_\mathrm{e}}\}$ for V_h, a unit tangent vector $\mathbf{t}_i$ is assigned to any edge E_i, $i = 1, \ldots, n_\mathrm{e}$, of T_h. Then, the global shape function associated to E_i is obtained by using local shape functions assigned to E_i, which are consistent with $\mathbf{t}_i$, see Figure 3.27. *The (global) Nédélec interpolation operator* is defined as

$$I_h^\mathrm{N} \boldsymbol{v} = \sum_{i=1}^{n_\mathrm{e}} \left(\int_{E_i} \boldsymbol{v} \cdot \mathbf{t}_i \, ds \right) \boldsymbol{\psi}_i.$$

The above discussion is valid for the 3D Nédélec finite element as well. Interpolating using the Nédélec finite element is straightforward in FEniCS. For example, to compute the Nédélec interpolant in the example of Section 3.4.2, we only need to replace the last 2 lines with the following lines:

```
# defining approximation space
Z_Ned = FunctionSpace(Mesh, "N1curl", 1)

# interpolating
Interpolant_Ned = interpolate(vf, Z_Ned)
```

3.4.4 AFFINE FAMILIES OF FINITE ELEMENTS

Similar to the mesh generation using a reference element discussed in Section 3.3, it is also useful in practice to generate a family of finite elements using *a reference finite element*. More specifically, let $(\hat{K}, \mathbb{P}_{\hat{K}}, \Sigma_{\hat{K}})$ be a simplicial Lagrange finite element with $\Sigma_{\hat{K}} = \{\hat{p}(\hat{b}_1), \ldots, \hat{p}(\hat{b}_N)\}$, where $\hat{b}_1, \ldots, \hat{b}_N$ are the nodes of the finite element. Let $f_K : \hat{K} \to K$ be an invertible affine mapping such that $K = f_K(\hat{K})$ is a simplex, see Figure 3.19. Then, it is not hard to show that $(K, \mathbb{P}_K, \Sigma_K)$ is also a finite element with

$$\mathbb{P}_K = \{p = \hat{p} \circ f_K^{-1}, \text{ for all } \hat{p} \in \mathbb{P}_{\hat{K}}\}, \text{ and } \Sigma_K = \{p(b_1), \ldots, p(b_N)\},$$

where $b_i = f_K(\hat{b}_i)$, $i = 1, \ldots, N$, are the nodes of K. Two finite finite elements $(\hat{K}, \mathbb{P}_{\hat{K}}, \Sigma_{\hat{K}})$ and $(K, \mathbb{P}_K, \Sigma_K)$ are then said to be *affine-equivalent*.

A family of finite elements $\{(K, \mathbb{P}_K, \Sigma_K)\}_{K \in T_h}$ associated to a mesh T_h is called *an affine family of finite elements* if all its finite elements are affine-equivalent to a finite element $(\hat{K}, \mathbb{P}_{\hat{K}}, \Sigma_{\hat{K}})$ called *the reference finite element*. If in addition, for any $K \in T_h$, the affine mapping $f_K : \hat{K} \to K$ belongs to $\mathbb{P}_{\hat{K}}$, then $\{(K, \mathbb{P}_K, \Sigma_K)\}_{K \in T_h}$ is called *an isoparametric family of finite elements.* By modifying the transformation of the local degrees of freedom of the reference element, it is also possible to extend the notion of affine families to the Raviart-Thomas and the Nédélec elements.

Affine families of finite elements are computationally beneficial in the sense that they allow one to reduce computations on all elements of a mesh to computations only on the reference element. Affine families of finite elements are also useful for theoretical studies. FEniCS automatically generates finite element spaces using a suitable reference finite element.

3.5 CONVERGENCE OF INTERPOLATIONS

Let $\{T_h\}_{h>0}$ be a sequence of successively refined meshes of a domain Ω such as the one shown in Figure 3.18. For each T_h, let I_h be the interpolation operator associated to a family of finite elements $\{(K, \mathbb{P}_K, \Sigma_K)\}_{K \in T_h}$. Then, given a function f, it is natural to ask if $I_h f$ converges to f as $h \to 0$. As mentioned in Section 3.1.1, normed linear spaces provide a suitable framework to answer this question in the sense that if f and $I_h f$ belong to a normed linear space $(X, \| \cdot \|)$, we can interpret convergence as

$$\lim_{h \to 0} \|f - I_h f\| = 0,$$

where $\|f - I_h f\|$ is *the error in the norm of* X. Convergence then generally depends on properties of f, the norm being used, properties of the domain Ω, and properties of the meshes $\{T_h\}_{h>0}$. In the following, we assume that Ω is a polygon or a polyhedron and that $\{T_h\}_{h>0}$ is *a shape-regular family of affine meshes*, where shape-regular means there is a constant c independent of h such that

$$\frac{h_K}{d_K} \leq c, \text{ for all } K \in T_h \text{ and all } T_h,$$

where h_K is the diameter of K and d_K is the diameter of the largest ball inscribed in K. One can show that $\frac{h_K}{d_K} \leq \frac{2}{\sin \alpha_K}$, where α_K is the smallest angle of K. Thus, $\{T_h\}_{h>0}$ cannot be shape-regular if elements of T_h become too "flat" in the sense that $\alpha_K \to 0$ as $h \to 0$, see Figure 3.28.

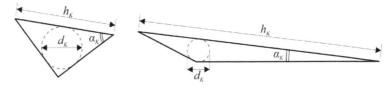

Figure 3.28 A family of meshes cannot be shape-regular if elements become too flat as $h \to 0$.

To study the convergence of the Lagrange interpolation operator I_h^k, we can use the normed linear spaces $(L^2(\Omega), \|\cdot\|_2)$ and $(H^1(\Omega), \|\cdot\|_{1,2})$ as the associated finite element spaces are H^1-conformal. In particular, suppose a function f is $s+1$ times differentiable with $1 \leq s \leq k$. Then, one can show that

$$\|f - I_h^k f\|_2 \leq C(f) h^{s+1},$$
$$\|f - I_h^k f\|_{1,2} \leq \tilde{C}(f) h^s, \tag{3.16}$$

where $C(f)$ and $\tilde{C}(f)$ do not depend on h. The above inequalities show that $\|f - I_h^k f\|_2$ and $\|f - I_h^k f\|_{1,2}$ tend to zero as $h \to 0$. The power of h in the right side of the above inequalities is called *the convergence rate*. Notice that the convergence rate depends on the smoothness of f and the norm being used. The maximum convergence rate of I_h^k is $k+1$ in the L^2-norm and k in the H^1-norm. The maximum convergence rate in the L^2-norm is called *the optimal convergence rate*.

Example 3.9. If f is not sufficiently differentiable, increasing the degree of polynomials k will not necessarily increase the convergence rate. For example, if f is only twice differentiable, the convergence rates of $I_h^1 f$ and $I_h^2 f$ will be 2 in the L^2-norm.

Let $\mathbf{v}$ be a 2D or a 3D vector field with twice differentiable components. Then, one can write

$$\|\mathbf{v} - I_h^{RT}\mathbf{v}\|_2 \leq \|\mathbf{v} - I_h^{RT}\mathbf{v}\|_d \leq D(\mathbf{v}) h,$$
$$\|\mathbf{v} - I_h^N\mathbf{v}\|_2 \leq \|\mathbf{v} - I_h^N\mathbf{v}\|_c \leq \tilde{D}(\mathbf{v}) h, \tag{3.17}$$

where $D(\mathbf{v})$ and $\tilde{D}(\mathbf{v})$ do not depend on h. Notice that since $I_h^{RT}\mathbf{v} \in H(\mathrm{div};\Omega)$ and $I_h^N\mathbf{v} \in H(\mathrm{curl};\Omega)$, the errors $\|\mathbf{v} - I_h^{RT}\mathbf{v}\|_d$ and $\|\mathbf{v} - I_h^N\mathbf{v}\|_c$ are well-defined. Therefore, the convergence rates of interpolations of twice differentialble vector fields using the Raviart-Thomas and the Nédélec finite elements (of degree 1) is 1 in the L^2, $H(\mathrm{div};\Omega)$, and $H(\mathrm{curl};\Omega)$ norms.

Following the program of Section 3.1.1 and using (3.6), it is straightforward to compute errors and convergence rates in FEniCS. As an example, the following code computes the convergence rates of various interpolations of the vector field

$$\mathbf{v}(x,y) = \left((x^2 + y^2)^b, xy\right), \tag{3.18}$$

with $b = 1.25$, on the unit square.

```
# defining the vector field
b_v = 1.25    # parameter of the vector field
vf = Expression((("pow(x[0]*x[0]+x[1]*x[1],b)", "x[0]*x[1]"),
                 b = b_v, degree = 5)

# number of divisions of meshes
Divisions = [3, 6, 9, 12]

# computing interpolants and errors
```

```
h = []  # max element sizes

# initializing errors
error_L2_Lag1, error_H1_Lag1 = [], []  # Lagrange type (1)
error_L2_Lag2, error_H1_Lag2 = [], []  # Lagrange type (2)
error_L2_RT, error_Hdiv = [], []        # RT
error_L2_Ned, error_Hcurl = [], []      # Nedelec

for n in Divisions:
    # defining the mesh
    mesh = UnitSquareMesh(n,n)

    # approximation spaces
    Z_Lag1 = VectorFunctionSpace(mesh, "Lagrange", 1)
    Z_Lag2 = VectorFunctionSpace(mesh, "Lagrange", 2)
    Z_RT = FunctionSpace(mesh, "RT", 1)
    Z_Ned = FunctionSpace(mesh, "N1curl", 1)

    # interpolants
    I_Lag1 = interpolate(vf, Z_Lag1)
    I_Lag2 = interpolate(vf, Z_Lag2)
    I_RT = interpolate(vf, Z_RT)
    I_Ned = interpolate(vf, Z_Ned)

    # max element size of the mesh
    h.append(mesh.hmax())

    # calculating errors
    error_L2_Lag1.append(errornorm(vf, I_Lag1, norm_type="L2"))
    error_H1_Lag1.append(errornorm(vf, I_Lag1, norm_type="H1"))

    error_L2_Lag2.append(errornorm(vf, I_Lag2, norm_type="L2"))
    error_H1_Lag2.append(errornorm(vf, I_Lag2, norm_type="H1"))

    error_L2_RT.append(errornorm(vf, I_RT, norm_type="L2"))
    error_Hdiv.append(errornorm(vf, I_RT, norm_type="Hdiv"))

    error_L2_Ned.append(errornorm(vf, I_Ned, norm_type="L2"))
    error_Hcurl.append(errornorm(vf, I_Ned, norm_type="Hcurl"))

# computing convergence rates
from math import log as ln  # log is a dolfin name too

# Lagrange of type 1
rate_L2_Lag1 = ln(error_L2_Lag1[-1]/error_L2_Lag1[-2])/ln(h[-1]/h[-2])
rate_H1_Lag1 = ln(error_H1_Lag1[-1]/error_H1_Lag1[-2])/ln(h[-1]/h[-2])

# Lagrange of type 2
```

```
rate_L2_Lag2 = ln(error_L2_Lag2[-1]/error_L2_Lag2[-2])/ln(h[-1]/h[-2])
rate_H1_Lag2 = ln(error_H1_Lag2[-1]/error_H1_Lag2[-2])/ln(h[-1]/h[-2])

# RT
rate_L2_RT = ln(error_L2_RT[-1]/error_L2_RT[-2])/ln(h[-1]/h[-2])
rate_Hdiv = ln(error_Hdiv[-1]/error_Hdiv[-2])/ln(h[-1]/h[-2])

# Nedelec
rate_L2_Ned = ln(error_L2_Ned[-1]/error_L2_Ned[-2])/ln(h[-1]/h[-2])
rate_Hcurl = ln(error_Hcurl[-1]/error_Hcurl[-2])/ln(h[-1]/h[-2])

print('Convergence Rates:')
print('Lagrange of type (1) => L2 = %.2f, H1   = %.2f'
      % (rate_L2_Lag1,rate_H1_Lag1))
print('Lagrange of type (2) => L2 = %.2f, H1   = %.2f'
      % (rate_L2_Lag2,rate_H1_Lag2))
print('Raviart-Thomas       => L2 = %.2f, Hdiv = %.2f'
      % (rate_L2_RT,rate_Hdiv))
print('Nedelec              => L2 = %.2f, Hcurl = %.2f'
      % (rate_L2_Ned,rate_Hcurl))
```

The above code works very similar to that of Section 3.1.1. A few comments are in order here. First, notice that following (3.18), the vector field vf is defined with a parameter b. This parameter is initialized as b = b_v, which is passed as an argument to Expression. After vf is defined, one can access the parameter b of vf by vf.b. For example, vf.b = 2.5, assigns the value 2.5 to b. Also notice that errornorm can compute various norms, which are specified by the variable norm_type. The output of the above code is:

```
Convergence Rates:
Lagrange of type (1) => L2 = 2.00, H1   = 1.00
Lagrange of type (2) => L2 = 2.99, H1   = 1.99
Raviart-Thomas       => L2 = 1.00, Hdiv  = 1.00
Nedelec              => L2 = 1.00, Hcurl = 1.00
```

These results are consistent with the theoretical values of convergence rates mentioned earlier.

From a purely mathematical point of view, higher convergence rates are preferable as they imply faster convergence for small h. In engineering applications, however, it may be computationally very expensive to consider very fine meshes and one may consider only a few meshes in the family $\{T_h\}$. In such cases where h is not necessarily "small" enough, the inequalities (3.16) and (3.17) suggest that errors also depend on the constants $C, \tilde{C}, D$, and $\tilde{D}$. For example, consider Figure 3.29 that shows error versus h diagrams for two cases. For $h = h_1, h_2$, the errors associated to the case with the higher convergence rate (the dashed curve) are larger than the ones associated to the lower rate case (the solid curve). Consequently, convergence rates may not be of direct practical interest. Nonetheless, as we will discuss in the following chapters,

computing convergence rates can be useful for debugging programs and for studying
the performance of numerical methods.

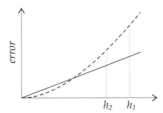

Figure 3.29   For meshes with not sufficiently small h, higher convergence rates do not necessarily imply smaller interpolation errors.

The approach that we considered in this section to study convergence by using a
sequence of successively refined meshes is called *the h-type approach*. Alternatively,
we may consider a fixed mesh and study the convergence by increasing the degree of
polynomials of finite elements. This is called *the p-type approach*. We only consider
the h-type approach in this book.

EXERCISES

Exercise 3.1. Show that $\{\psi_0, \psi_1, \ldots, \psi_N\}$ is a basis for $\mathbb{P}_1(T_h)$, where the functions
ψ_i are defined in (3.2).

Exercise 3.2. Given a finite element $(K, \mathbb{P}, \Sigma)$, show that if $p \in \mathbb{P}$, then $I_K p = p$, that
is, the local interpolation operator I_K does not change members of $\mathbb{P}$.

Exercise 3.3. Verify the barycentric coordinates of the unit 2- and 3-simplices given
in Examples 3.1 and 3.2.

Exercise 3.4. Let K be a tetrahedron. Show that any $p \in \mathbb{P}_1(K)$ can be uniquely
determined by its values at the vertices of K.

Exercise 3.5. For $n = 1, 2, 3$, and $k = 1$, verify that the definition of n-simplex of
type (k) based on (3.11) results in n-simplex of type (1) introduced in Sections 3.1.2
and 3.2.1.

Exercise 3.6. Show that local shape functions of 1- and 2-simplices of type 2 are
those given in Examples 3.3 and 3.4.

Exercise 3.7. Write the barycentric coordinates of the nodes of 3-simplex of type
(2) and derive the associated local shape functions. The conventional finite element
diagram of this element is shown in Figure 3.9.

Exercise 3.8. *The 2D Crouzeix-Raviart finite element* is a Lagrange finite element
$(K, \mathbb{P}_1(K), \Sigma_K)$, where K is a 2-simplex and the nodes of this element have the

barycentric coordinates $b_0 = (0, 1/2, 1/2)$, $b_1 = (1/2, 0, 1/2)$, and $b_2 = (1/2, 1/2, 0)$. Plot the conventional finite element diagram of this element and derive its local shape functions in terms of barycentric coordinates. Is this element equal to 2-simplex of type (1)?

Exercise 3.9. Show that the polynomials given in (3.12) are the local shape functions of the Hermite 1-simplex of type (3).

Exercise 3.10. Show that $\mathbb{RT}$, the underlying polynomial space of the Raviart-Thomas finite element, is a linear space with $\dim \mathbb{RT} = 3$.

Exercise 3.11. In Example 3.5, show that $\sigma_1(\boldsymbol{\theta}_0) = \sigma_2(\boldsymbol{\theta}_0) = 0$.

Exercise 3.12. Show that $\mathbb{NE}$, the underlying polynomial space of the Nédélec finite element, is a linear space with $\dim \mathbb{NE} = 3$.

Exercise 3.13. For the Nédélec finite element, show that $\sigma_i(\boldsymbol{\theta})_j = \delta_{ij}$, where the degrees of freedom σ_i and the local shape functions $\boldsymbol{\theta}_i$, $i = 1, 2, 3$, are introduced in Section 3.2.4.

Exercise 3.14. Suppose I_h is the global interpolation operator associated to a mesh T_h and the finite element space V_h. Show that if $f \in V_h$, then $I_h f = f$.

Exercise 3.15. Show that the set of the functions $\{\psi_1, \dots, \psi_{n_v + n_e}\}$ introduced in Example 3.7 is a basis for $\mathbb{P}_2(T_h)$.

Exercise 3.16. Determine the dimension of the finite element space associated to 3-simplices of type (k) on a 3D mesh with $k = 2$ and $k = 3$.

Exercise 3.17. Determine the dimension of the finite element space on a 2D mesh associated to a family of 2D Crouzeix-Raviart finite elements introduced in Exercise 3.8. Is this finite element space equal to the finite element space associated to a family of 2-simplices of type (1)?

Exercise 3.18. Determine the dimension of the finite element spaces associated to a family of Hermite 2- and 3-simplices of type (3).

COMPUTER EXERCISES

Computer Exercise 3.1. By modifying the program of Section 3.1.1, approximate the L^2 and H^1 convergence rates of Largange interpolants of type (1) for the functions (a) $f(x) = x^{0.8}$, and (b) $f(x) = x^{0.2}$. Justify your results in case convergence rates are not optimal.

Computer Exercise 3.2. Plot local shape functions of 1-simplex of type (3) on the unit 1-simplex $K = [0, 1]$.

Computer Exercise 3.3. Plot local shape functions of 2-simplex of type (2) on the unit 2-simplex.

Computer Exercise 3.4. Plot the local shape functions of the 2D Raviart-Thomas finite element on a 2-simplex with the vertices $(0,0)$, $(2,1)$, and $(-1,3)$.

Computer Exercise 3.5. Plot the local shape functions of the 2D Nédélec finite element on a 2-simplex with the vertices $(0,0)$, $(2,1)$, and $(-1,3)$.

Computer Exercise 3.6. Let Ω be the unit square with the vertices $v_1 = (0,0)$, $v_2 = (1,0)$, $v_3 = (1,1)$, and $v_4 = (0,1)$, and let $T_h = \{K_1, K_2\}$ be a mesh of Ω, where K_1 and K_2 are triangles with the vertices $\{v_1, v_2, v_3\}$ and $\{v_1, v_3, v_4\}$, respectively. Plot the vector field $\boldsymbol{p}(x,y) = (x^2 + 2xy, x + y^2)$ on Ω. Also separately plot the finite element interpolants $I_h^1 \boldsymbol{p}$, $I_h^2 \boldsymbol{p}$, $I_h^{RT} \boldsymbol{p}$, and $I_h^N \boldsymbol{p}$. Which of these interpolants are discontinuous? Which one coincides with $\boldsymbol{p}$?

Computer Exercise 3.7. Determine the dimension of the finite element space on a 1D mesh associated to 1-simplices of type (2). Plot the corresponding locally supported global shape functions on a simple mesh consisting of 3 elements.

Computer Exercise 3.8. Consider a uniform mesh of $\Omega = (0,4)$ with vertices $\{0,1,2,3,4\}$. Plot global shape functions associated to the vertex 2 induced by a family of: (a) 1-simplices of type (3), and (b) Hermite 1-simplices of type (3). Compare these global shape functions.

Computer Exercise 3.9. Consider a sequence of successively refined meshes for the unit square of Computer Exercise 3.6 and plot the errors $\|v - I_h^1 v\|_2$, $\|v - I_h^2 v\|_2$, $\|v - I_h^{RT} v\|_2$, and $\|v - I_h^N v\|_2$ versus h, where $v(x,y) = ((x^2 + y^2)^{\frac{b}{2}}, \sin(xy))$, with $b = 2.5$. Determine the convergence rate for each case.

Computer Exercise 3.10. Repeat Computer Exercise 3.9 with $b = 1.5$, and $b = 0.5$. Do the convergence rates coincide with those of Computer Exercise 3.9? Why?

Computer Exercise 3.11. Let Ω be the unit cube with the vertices $v_1 = (0,0,0)$, $v_2 = (1,0,0)$, $v_3 = (1,1,0)$, $v_4 = (0,1,0)$, $v_5 = (0,0,1)$, $v_6 = (1,0,1)$, $v_7 = (1,1,1)$, and $v_8 = (0,1,1)$. Plot the errors $\|v - I_h^1 v\|_2$, $\|v - I_h^2 v\|_2$, $\|v - I_h^{RT} v\|_2$, and $\|v - I_h^N v\|_2$ versus h, where $v(x,y,z) = ((x^2 + y^2 + z^2)^{\frac{b}{2}}, \sin(xyz), \cos(xyz))$, with $b = 2.5$. Determine the convergence rate for each case.

Computer Exercise 3.12. Repeat Computer Exercise 3.11 with $b = 1.5$, and $b = 0.5$. Do the convergence rates coincide with those of Computer Exercise 3.11?

COMMENTS AND REFERENCES

Instructions for installing FEniCS together with several tutorial and solved examples are avalible in the FEniCS Project webpage https://fenicsproject.org/, see also Appendix A. The FEniCS Tutorial [9] is a good guide for beginners. A more complete tutorial is the FEniCS Book [10], though some parts of this tutorial are outdated. A good introduction to Python programming for scientific computations is given by [8], also see Appendix B.

The definition of a finite element given in Section 3.2 was first introduced in the classic book of Ciarlet [5]. In this book, a complete study of several types of finite elements that are commonly used in engineering and science is avaiable. It also discusses curved-face elements and additional numerical errors that are induced by approximating curved boundaries. Another good reference for common finite elements including the Raviart-Thomas and the Nédélec elements is [6]. Proofs of the error estimates (3.16) and (3.17) are also available in [6, Chapter 1].

4 Conforming Finite Element Methods for PDEs

In this chapter, we will study finite element approximations of solutions of time-independent and time-dependent classes of second-order partial differential equations (PDEs) that arise in many engineering and scientific applications. After defining the general form of these PDEs, we introduce the notion of weak forms for boundary value problems associated to these PDEs. We then employ finite element spaces introduced in the previous chapter to obtain finite element methods for approximating solutions of weak forms. By considering several examples, we also discuss implementation of PDEs subject to different boundary conditions in FEniCS. Finally, a brief introduction to mixed finite element methods and their implementation in FEniCS is provided.

Approximation methods of this chapter are called *conforming* finite element methods in the sense that finite element spaces containing approximate solutions are subsets of the Sobolev spaces that contain weak solutions of PDEs.

4.1 SECOND-ORDER ELLIPTIC PDES

Let Ω be an open domain in $\mathbb{R}^n$, $n = 1, 2, 3$, and let $\partial\Omega$ be its boundary. We consider the PDE

$$-\text{div}(\nabla u \cdot \mathbb{D}) + \boldsymbol{b} \cdot \nabla u + cu = f \text{ in } \Omega, \tag{4.1}$$

where $u : \Omega \to \mathbb{R}$ is the unknown function, $\mathbb{D}(x)$ is an $n \times n$ matrix at any point $x = (x_1, \ldots, x_n)$ of Ω with the components $d_{ij}(x)$, and $\boldsymbol{b}(x) = (b_1(x), \ldots, b_n(x))$ is a vector field. The components d_{ij} and b_i together with c and f are assumed to be known real-valued functions on Ω. The PDE (4.1) can be written more explicitly as

$$-\sum_{i,j=1}^{n} \partial_{x_j}(d_{ij}\,\partial_{x_i}u) + \sum_{i=1}^{n} b_i \partial_{x_i} u + cu = f \text{ in } \Omega,$$

and therefore, it is a second-order PDE, that is, the highest order derivative in (4.1) is 2. A PDE of the form (4.1) is called *elliptic* if for all row vectors $\mathbf{w} = (w_1, \ldots, w_n) \in \mathbb{R}^n$, there exists $\beta > 0$ such that

$$\mathbf{w} \cdot \mathbb{D}(x) \cdot \mathbf{w}^T \geq \beta \|\mathbf{w}\|^2, \text{ at any } x \in \Omega, \tag{4.2}$$

where $\mathbf{w}^T$ is the transpose of $\mathbf{w}$ and $\|\mathbf{w}\| = \sqrt{\mathbf{w} \cdot \mathbf{w}^T}$ is the standard norm of $\mathbf{w}$ in $\mathbb{R}^n$. Equivalently, (4.2) can be stated as

$$\sum_{i,j=1}^{n} d_{ij}(x)\,w_i w_j \geq \beta \sum_{i=1}^{n} w_i^2, \text{ at any } x \in \Omega.$$

Elliptic PDEs of the form (4.1) arise in many applications. A few examples include:

- For $n = 1$, $\boldsymbol{b} = 0$, $c = 0$, and $\mathbb{D}(x) = E(x)A(x)$, one obtains the equilibrium equation $\partial_x(EA\partial_x u) + f = 0$, of a 1D linearly elastic bar in terms of the displacement $u(x)$, where $A(x)$ and $E(x)$ are respectively the cross section area and Young's modulus of the bar.
- With $\boldsymbol{b} = 0$, $c = 0$, and $\mathbb{D}$ being the identity matrix $\mathbb{I}$, one recovers Poisson's equation $-\Delta u = f$. This equation has several applications such as torsion of prismatic rods, deflections of elastic membranes, incompressible potential flows, and electrostatics.
- For $\mathbb{D} = k\mathbb{I}$, and $c = 0$, with $k > 0$, one obtains the heat transfer equation.
- For $\mathbb{D} = k\mathbb{I}$, with $k > 0$, (4.1) reduces to the advection-diffusion equation.

The PDE (4.1) has infinitely many solutions in general. To solve this PDE for a specific solution, one also needs to specify suitable *boundary conditions* to obtain a well-defined *boundary value problem*. We will discuss some possible choices of boundary conditions later in this chapter.

4.2   WEAK FORMULATIONS OF ELLIPTIC PDES

For obtaining a solution of the PDE (4.1), we should supplement it with a suitable boundary condition as well. Solving the resulting boundary value problem means finding a function $u : \Omega \to \mathbb{R}$ that satisfies the PDE (4.1) in Ω and also satisfies the boundary condition on the boundary $\partial\Omega$. Thus, the solution process involves three basic aspects: Finding a unique function (*existence* and *uniqueness*) that is twice differentiable (*regularity*) and satisfies the boundary value problem.

It was observed that the simultaneous achievement of the above three aspects is hard. A common approach is to first study the existence and the uniqueness of solutions and then their regularity. To this end, alternative formulations of boundary value problems called *weak formulations* are employed. Usually, it is theoretically and computationally more convenient to study solutions of weak formulations, which are also called *weak solutions*. If a weak solution is sufficiently differentiable (regularity), then it is also a solution of the associated boundary value problem. Therefore, weak solutions are good candidates for solutions of boundary value problems. It should be mentioned that in some applications such as continuum mechanics, weak formulations rather than PDEs are considered to represent the real physics of systems.

In the remainder of this section, we write the weak formulation of some boundary value problems associated to (4.1). We do not discuss regularity of weak solutions in this book. Under some mild conditions, one can show that weak solutions of elliptic PDEs are also sufficiently differentiable. Studying the regularity of weak solutions of more general PDEs is usually a challenging task.

4.2.1   DIRICHLET BOUNDARY CONDITION

The simplest boundary condition for (4.1) is *the homogeneous Dirichlet boundary condition*

$$u = 0, \text{ on } \partial\Omega. \tag{4.3}$$

Let v be an arbitrary function that vanishes on the boundary $\partial\Omega$. By multiplying (4.1) by v and integrating over Ω, one obtains

$$-\int_\Omega v\,\mathrm{div}(\nabla u \cdot \mathbb{D}) + \int_\Omega (\boldsymbol{b}\cdot\nabla u)v + \int_\Omega cuv = \int_\Omega fv. \tag{4.4}$$

Green's formula (2.4) implies that

$$-\int_\Omega v\,\mathrm{div}(\nabla u \cdot \mathbb{D}) = \int_\Omega \nabla u \cdot \mathbb{D} \cdot (\nabla v)^T - \int_{\partial\Omega} (v\nabla u \cdot \mathbb{D}) \cdot \boldsymbol{n}, \tag{4.5}$$

where $(\nabla v)^T$ is the transpose of the row vector field ∇v and $\boldsymbol{n}$ is the outward unit normal vector field at $\partial\Omega$. Using the above relation and $v|\partial\Omega = 0$, we can write (4.4) as

$$\int_\Omega \nabla u \cdot \mathbb{D} \cdot (\nabla v)^T + \int_\Omega (\boldsymbol{b}\cdot\nabla u)v + \int_\Omega cuv = \int_\Omega fv. \tag{4.6}$$

If u is a solution of (4.1) subject to (4.3), then it satisfies (4.6). Both sides of (4.6) are well-defined for $u, v \in H_0^1(\Omega)$, see Example 2.11. Therefore, we can define *the following weak formulation* of (4.1) subject to the Dirichlet boundary condition (4.3): Find $u \in H_0^1(\Omega)$ such that

$$B(u, v) = \int_\Omega fv, \text{ for all } v \in H_0^1(\Omega), \tag{4.7}$$

where the bilinear form $B : H_0^1(\Omega) \times H_0^1(\Omega) \to \mathbb{R}$ is given by

$$
\begin{aligned}
B(u, v) &= \int_\Omega \left\{ \nabla u \cdot \mathbb{D} \cdot (\nabla v)^T + (\boldsymbol{b}\cdot\nabla u)v + cuv \right\} \\
&= \int_\Omega \left\{ \sum_{i,j=1}^n d_{ij}\partial_{x_i} u\,\partial_{x_j} v + v \sum_{i=1}^n b_i \partial_{x_i} u + cuv \right\}.
\end{aligned}
\tag{4.8}
$$

Notice that (4.1) involves second-order derivatives of u, however, the weak formulation (4.7) only involves first-order derivatives of u and v. A solution of the weak formulation is called *a weak solution* of the PDE (4.1) subject to (4.3). To show that a weak solution is also a solution of the PDE (also called *a strong solution*), one should verify that the weak solution is twice differentiable (regularity). Hence, a strong solution is also a weak solution but the converse may not hold, in general.

The arbitrary function v in the weak formulation (4.7) is called *the test function*. The space that contains weak solutions is called *the solution space* or *the trial space*. The space that contains test functions is called *the test space*. The solution and the test spaces of (4.7) are the same and are equal to $H_0^1(\Omega)$.

In (4.7), the homogeneous Dirichlet boundary condition (4.3) is directly imposed in the solution space $H_0^1(\Omega)$. This boundary condition is then called *an essential boundary condition* for (4.7).

One may wonder why in (4.7) we assumed that $v|_{\partial\Omega} = 0$. If $v|_{\partial\Omega} \neq 0$, then the term $\int_{\partial\Omega} (v\nabla u \cdot \mathbb{D}) \cdot \boldsymbol{n}$ on the right side of (4.5) that contains first-order derivatives of u should be added to the bilinear form $B(u, v)$ as well. One can show that if $u \in H^1(\Omega)$, then u is integrable on the boundary. However, the first-order derivatives of u are not integrable on the boundary, in general. Consequently, the bilinear form $B(u, v)$ is not well-defined as a function $H^1(\Omega) \times H^1(\Omega) \to \mathbb{R}$, if $v|_{\partial\Omega} \neq 0$; See also the discussion of Section 2.8 regarding Green's formulas in terms of Sobolev spaces.

A *non-homogeneous* Dirichlet boundary condition for (4.1) reads

$$u = g, \text{ on } \partial\Omega, \tag{4.9}$$

where $g : \Omega \to \mathbb{R}$ is a known function. Notice that we assumed g is defined on the whole domain and not only on its boundary. If u is a solution of (4.1) subject to (4.9), then $w = u - g$ vanishes at the boundary and is a solution of the problem: Find $w \in H_0^1(\Omega)$ such that

$$B(w, v) = \int_\Omega fv - B(g, v), \text{ for all } v \in H_0^1(\Omega), \tag{4.10}$$

where the bilinear form B is defined in (4.8). Thus, the solution of problems subject to non-homogeneous Dirichlet boundary conditions can be reduced to the solution of problems subject to the homogeneous Dirichlet boundary condition. Alternatively, one may consider the following weak formulation for non-homogeneous Dirichlet boundary conditions: Find $u \in H^1(\Omega)$ satisfying $u|_{\partial\Omega} = g$ such that

$$B(u, v) = \int_\Omega fv, \text{ for all } v \in H_0^1(\Omega). \tag{4.11}$$

The trial and the test spaces of the second weak formulation are not the same. This weak formulation is more suitable in case the function g is only defined at the boundary.

4.2.2 NEUMANN BOUNDARY CONDITION

A Neumann boundary condition for (4.1) is given by

$$(\nabla u \cdot \mathbb{D}) \cdot \boldsymbol{n} = g, \text{ on } \partial\Omega, \tag{4.12}$$

where $g : \partial\Omega \to \mathbb{R}$ is a known function on the boundary. Notice that if $\mathbb{D}$ is the identity matrix $\mathbb{I}$, then (4.12) simply states that the normal derivative $\partial_{\boldsymbol{n}} u$ of u at $\partial\Omega$ is g.

By following the approach used for deriving (4.7), one obtains the following weak formulation for the PDE (4.1) subject to (4.12): Find $u \in H^1(\Omega)$ such that

$$B(u, v) = \int_\Omega fv + \int_{\partial\Omega} gv, \text{ for all } v \in H^1(\Omega), \tag{4.13}$$

where the bilinear form $B(u,v)$ is the same as (4.8). The solution and the test space of this weak formulation are $H^1(\Omega)$. The derivation of (4.13) is left as an exercise. Notice that the Neumann boundary condition (4.12) is implicitly imposed by the weak formulation and not directly by the solution space. This boundary condition is then called *a natural boundary condition* for (4.13).

It is also possible to have a combination of Dirichlet and Neumann boundary conditions as follows: Suppose that $\partial\Omega$ consists of the two parts Γ_D and Γ_N and consider *the mixed Dirichlet-Neumann boundary condition*

$$\begin{cases} u = \hat{g}, & \text{on } \Gamma_D, \\ (\nabla u \cdot \mathbb{D}) \cdot \boldsymbol{n} = g, & \text{on } \Gamma_N. \end{cases} \tag{4.14}$$

Let

$$H^1_D(\Omega) = \{v \in H^1(\Omega) : \ v = 0 \text{ on } \Gamma_D\}.$$

Then, a weak formulation for the PDE (4.1) subject to (4.14) can be stated as: Find $u \in H^1(\Omega)$ such that $u|_{\Gamma_D} = \hat{g}$ and

$$B(u,v) = \int_\Omega fv + \int_{\Gamma_N} gv, \text{ for all } v \in H^1_D(\Omega), \tag{4.15}$$

where the bilinear form $B(u,v)$ is given by (4.8). The derivation of (4.15) is left as an exercise.

4.2.3   ROBIN BOUNDARY CONDITION

Given two functions $r : \partial\Omega \to \mathbb{R}$ and $g : \partial\Omega \to \mathbb{R}$, *a Robin boundary condition* for the PDE (4.1) is given by

$$ru + (\nabla u \cdot \mathbb{D}) \cdot \boldsymbol{n} = g, \text{ on } \partial\Omega. \tag{4.16}$$

A weak formulation for (4.1) subject to the Robin boundary condition (4.16) can be stated as: Find $u \in H^1(\Omega)$ such that

$$B(u,v) + \int_{\partial\Omega} ruv = \int_\Omega fv + \int_{\partial\Omega} gv, \text{ for all } v \in H^1(\Omega), \tag{4.17}$$

where the bilinear form $B(u,v)$ is defined in (4.8). The derivation of (4.17) is left as an exercise. Notice that similar to Neumann boundary conditions, Robin boundary conditions are imposed as a natural boundary condition in (4.17).

Let us emphasize that the selection of a specific boundary condition depends on the physical problem that is modeled by (4.1). For example, to model a membrane fixed at its boundary similar to the one discussed in Chapter 1, one can employ the homogeneous Dirichlet boundary condition. It is also possible to have more than one possible choices of boundary conditions for some physical problems.

4.3 WELL-POSEDNESS OF WEAK FORMULATIONS

The weak formulations introduced in Section 4.2 can be considered as special cases of *the following abstract problem*: Let $(X, \| \cdot \|)$ be a normed linear space and suppose $A : X \times X \to \mathbb{R}$ and $F : X \to \mathbb{R}$ are respectively a bilinear form and a linear functional. We consider the abstract problem: Find $u \in X$ such that

$$A(u,v) = F(v), \text{ for all } v \in X. \tag{4.18}$$

The data X, A, and F for the weak formulations of Section 4.2 are the followings:

- The formulation (4.7) for the homogeneous Dirichlet boundary condition:

$$X = H_0^1(\Omega), \quad A(u,v) = B(u,v), \quad F(v) = \int_\Omega fv;$$

- The formulation (4.13) for the Neumann boundary condition:

$$X = H^1(\Omega), \quad A(u,v) = B(u,v), \quad F(v) = \int_\Omega fv + \int_{\partial\Omega} gv;$$

- The formulation (4.15) for the mixed Dirichlet-Neumann boundary condition with $\hat{g} = 0$:

$$X = H_D^1(\Omega), \quad A(u,v) = B(u,v), \quad F(v) = \int_\Omega fv + \int_{\Gamma_N} gv;$$

- The formulation (4.17) for the Robin boundary condition:

$$X = H^1(\Omega), \quad A(u,v) = B(u,v) + \int_{\partial\Omega} ruv, \quad F(v) = \int_\Omega fv + \int_{\partial\Omega} gv.$$

By inspecting the derivations of the above weak formulations, we observe that linearity of $A(u,v)$ with respect to u and v stems from different origins: The linearity with respect to u follows form the linearity of the left side of the PDE (4.1) with respect to u while the linearity with respect to v simply follows from the derivation of these weak formulations. This means that if we replace (4.1) with a nonlinear PDE and follow the derivation of Section 4.2, we will obtain a weak formulation of the form $A(u,v) = F(v)$, where $A(u,v)$ is only linear with respect to its second argument v.

Our main goal in this section is to study solvability of the abstract problem (4.18). For many problems in engineering and science, it is reasonable to expect that (4.18) has only one solution u for each data F. It is also reasonable to assume the solution u continuously depends on the data F in the sense that "small" changes in F will result in "small" changes in the corresponding solution u. This continuity condition can be stated as: There is $\alpha > 0$, such that for any F we have

$$\|u\| \le \alpha \|F\|, \tag{4.19}$$

where u is the solution associated to F and $\| \cdot \|$ denotes appropriate norms. We say the abstract problem (4.18) is *well-posed* (in the Hadamard sense) if it has a unique solution and if the continuity condition (4.19) holds. Notice that the above definition of well-posedness is not appropriate for some applications including those with non-unique solutions such as buckling of columns.

The Lax-Milgram lemma provides a sufficient condition for the well-posedness of the problem (4.18): Suppose A is X-*elliptic*, also called *coercive*, in the sense that there exists $\gamma > 0$ such that

$$A(v,v) \geq \gamma \|v\|^2, \text{ for all } v \in X. \tag{4.20}$$

Then, the Lax-Milgram lemma states that the problem (4.18) is well-posed with $\alpha = 1/\gamma$ in (4.19). Notice that coercivity is only a sufficient condition for well-posedness and it is possible to have well-posed non-coercive problems as well.

One can show that, under some suitable conditions, all the weak formulations of (4.1) with the boundary conditions discussed in Section 4.2 are coercive and consequently, are well-posed. Other important examples of coercive problems that do not belong to the class of elliptic PDEs of the form (4.1) include linearized elasticity and the biharmonic problem. Later we will see that a very useful property of coercive problems is that *well-posedness of these problems is automatically inherited by their discretizations.*

4.4 VARIATIONAL STRUCTURE

It is possible that the abstract problem (4.18) admits *a variational structure* in the sense that it corresponds to the minimization of a functional. More specifically, suppose $A(u,v)$ is symmetric, that is, $A(u,v) = A(v,u)$, and consider the functional $J : X \to \mathbb{R}$ given by

$$J(v) = \frac{1}{2}A(v,v) - F(v). \tag{4.21}$$

For any $t \in \mathbb{R}$ and $u, v \in X$, we can write

$$J(u+tv) - J(u) = t\left[A(u,v) - F(v)\right] + \frac{t^2}{2}A(v,v). \tag{4.22}$$

Therefore, if u solves the abstract problem (4.18) and if A is positive, that is, $A(v,v) \geq 0$, for all $v \in X$, we have $J(u+tv) - J(u) \geq 0$.

In summary, we conclude that if A is *symmetric and positive*, then a solution of the problem (4.18) is a minimizer of J. Conversely, it is not hard to show that a minimizer of J is a solution of the problem (4.18). We leave the proof as an exercise. When the abstract problem (4.18) corresponds to a minimizer of the functional (4.21), it is also called *a variational formulation.*

Notice that if the bilinear form A is coercive, then it is positive. *Thus, the Lax-Milgram lemma implies that if the bilinear form A is symmetric and coercive, then the unique solution of (4.18) is the unique minimizer of (4.21).* In practice, the functional (4.21) usually represents a specific type of energy. For example, it represents the stored energy in linearized elasticity.

Inspection of the bilinear form $B(u,v)$ defined in (4.8) suggests that $B(u,v)$ is symmetric if the vector field $\boldsymbol{b}$ of the PDE (4.1) vanishes. In this case, it is straightforward to show that the weak formulations of Section 4.2 are also variational formulations. The verification of this result is left as an exercise.

4.5   THE GALERKIN METHOD AND FINITE ELEMENT METHODS

The Galerkin method provides a framework for approximating solutions of the abstract problem (4.18). Assume X_h is a finite-dimensional linear subspace of possibly infinite-dimensional, linear space X. To approximate a solution of (4.18) using the Galerkin method, we *discretize* the problem (4.18) as: Find $u_h \in X_h$ such that

$$A(u_h, v_h) = F(v_h), \text{ for all } v_h \in X_h. \tag{4.23}$$

Here, u_h is the *approximate* solution, v_h is the test function, and X_h is both the solution (or trial) space and the test space. The approximation scheme (4.23) is called *a conforming method* as $X_h \subset X$. If $X_h \not\subset X$, it is called *a non-conforming method*.

If the finite-dimensional linear space X_h is a finite element space and $X_h \subset X$, the problem (4.23) is called *a conforming finite element method* for approximating a solution of (4.18).

As discussed in Section 4.4, if the bilinear form $A(u,v)$ of the abstract problem (4.18) is symmetric, then a solution of (4.18) is also a minimizer of the functional $J(v)$ in X, where J is defined in (4.21). In this case, a solution of the discrete problem (4.23) is a minimizer of J in the space X_h. Approximating a minimizer of J in X by using a minimizer of J in X_h is called *the Ritz method*. In the sense discussed above, the Ritz method is equivalent to the Galerkin method.

To obtain conforming finite element methods for elliptic PDEs subject to the boundary conditions of Section 4.2, let T_h be a simplicial mesh of the underlying domain Ω and let V_h be the finite element space associated to a family of simplicial Lagrange element of type (k) on T_h. Recall that in Section 3.4.1.1, we showed that $V_h \subset H^1(\Omega)$.

By using suitable data mentioned in Section 4.3 for different boundary conditions, (4.23) leads to conforming finite element methods for boundary value problems associated to the PDE (4.1), where X_h is selected as follows:

– The homogeneous Dirichlet boundary condition:

$$X_h = \{v_h \in V_h : v_h = 0 \text{ on } \partial\Omega\}; \tag{4.24}$$

– Mixed Dirichlet-Neumann boundary condition:

$$X_h = \{v_h \in V_h : v_h = 0 \text{ on } \Gamma_D\}; \tag{4.25}$$

– Neumann and Robin boundary conditions: $X_h = V_h$.

In the remainder of this section, we study well-posedness and convergence of conforming finite element methods associated to (4.23).

4.5.1 THE STIFFNESS MATRIX

We begin by showing that solving (4.23) is equivalent to solving a system of linear equations. Let $\{\psi_1, \ldots, \psi_N\}$ be a basis for the finite-dimensional space X_h with $\dim X_h = N$, where X_h is not necessarily a finite element space. Since $u_h \in X_h$, we can write

$$u_h = \sum_{j=1}^{N} U_j \psi_j, \tag{4.26}$$

where $U_1, \ldots, U_N$ are unknown constants to be determined. It is straightforward to show that u_h solves (4.23) if and only if

$$A(u_h, \psi_i) = F(\psi_i), \quad i = 1, \ldots, N.$$

By using (4.26), the above equations can be written as the linear system

$$[A]_{N \times N} \cdot \mathbb{U}_{N \times 1} = \mathbb{F}_{N \times 1}, \tag{4.27}$$

where $[A]$ is called *the stiffness matrix* with the components $[A]_{ij} = A(\psi_j, \psi_i)$, and $\mathbb{U}$ and $\mathbb{F}$ are vectors given by

$$\mathbb{U} = \begin{bmatrix} U_1 \\ \vdots \\ U_N \end{bmatrix}, \quad \mathbb{F} = \begin{bmatrix} F(\psi_1) \\ \vdots \\ F(\psi_N) \end{bmatrix}.$$

Notice that the stiffness matrix $[A]$ is the matrix representation of the bilinear form $A : X_h \times X_h \to \mathbb{R}$ in the basis $\{\psi_1, \ldots, \psi_N\}$ in the sense discussed in Section 2.5. It is straightforward to verify that if A is a symmetric bilinear form, then $[A]$ will be a symmetric matrix.

If X_h is a finite element space, then $\{\psi_1, \ldots, \psi_N\}$ can be chosen to be the locally supported global shape functions of X_h. If, as for boundary value problems associated to elliptic PDEs, $A(u, v)$ is obtained by integrating terms that involve the multiplication of u and v or their derivatives, the selection of global shape functions will lead to *sparse* stiffness matrices, that is, most entries of $[A]$ will be zero. The sparsity of $[A]$ is very useful for efficiently computing the solution of the linear system (4.27).

Example 4.1. Consider the mesh of a square with 8 elements shown in the middle of Figure 3.18. Suppose we use the finite element space associated to a family of 1-simplex of type (1) to obtain finite element methods for an elliptic PDE. Then, the size of the stiffness matrix $[A]$ will depend on the selected boundary condition. In particular, the size of $[A]$ is 1×1 for the homogeneous Dirichlet boundary condition, 6×6 for a mixed Dirichlet-Neumann boundary condition with Γ_D being the bottom edge of the square, and 9×9 for Neumann and Robin boundary conditions.

Example 4.2. Suppose the mesh of Figure 3.23 is employed to obtain a finite element method for an elliptic PDE using 2-simplex of type (2). Then, the entry of the stiffness matrix corresponding to the global shape functions shown on the left and the right of Figure 3.23 is zero.

4.5.2   WELL-POSEDNESS OF COERCIVE DISCRETE PROBLEMS

Generally speaking, the well-posedness of the abstract problem (4.18) does not imply the well-posedness of the discrete problem (4.23). *Coercive problems are exceptions in this regard.* More specifically, suppose the coercivity condition (4.20) holds and let $X_h \subset X$. Then, we can write

$$A(v_h, v_h) \geq \gamma \|v_h\|^2, \text{ for all } v_h \in X_h, \tag{4.28}$$

and therefore, $A : X_h \times X_h \to \mathbb{R}$ is also coercive. The Lax-Milgram lemma then tells us that the discrete problem (4.23) is well-posed and has a unique solution. Thus, the well-posedness of conformal discretizations of coercive problems automatically follows from the well-posedness of these problems.

The coercivity of $A : X_h \times X_h \to \mathbb{R}$ implies that the stiffness matrix $[A]_{N \times N}$ is *positive definite* in the sense that

$$\mathbf{v}^T \cdot [A] \cdot \mathbf{v} \geq 0, \text{ for all column vectors } \mathbf{v} \in \mathbb{R}^N. \tag{4.29}$$

Symmetric, positive definite matrices have positive eigenvalues.

4.5.3   CONVERGENCE OF FINITE ELEMENT SOLUTIONS

Let $\{T_h\}$ be a family of meshes of a domain Ω. For each mesh T_h of this family, one can write a finite element method in the form (4.23) with the solution $u_h \in X_h$. Consequently, we obtain a family of approximate solutions $\{u_h\}$ associated to $\{T_h\}$. We say this family of finite element solutions *converges* in the norm $\| \cdot \|$ to a solution $u \in X$ of the problem (4.18) if

$$\lim_{h \to 0} \|u - u_h\| = 0,$$

where $\|u - u_h\|$ is *the error in the norm* $\| \cdot \|$. Notice that since $X_h \subset X$, we have $u - u_h \in X$ and thus, $\|u - u_h\|$ is well-defined when $\| \cdot \|$ is the norm of X. As will be discussed later, it is also possible to study convergence in a norm other than the norm of the solution space X as long as the other norm of $u - u_h$ is also well-defined.

To study the convergence of finite element solutions, let us assume that the bilinear form $A : X \times X \to \mathbb{R}$ is coercive and continuous. This means there are two positive constants γ and ξ such that for all $v, w \in X$, we have

$$A(v, v) \geq \gamma \|v\|^2, \quad \text{and} \quad |A(v, w)| \leq \xi \|v\| \|w\|.$$

The coercivity of A implies that the problem (4.18) and the discrete problem (4.23) are well-posed and have unique solutions u and u_h, respectively. Since for any $w_h \in X_h$, we have $A(u, w_h) = F(w_h) = A(u_h, w_h)$, one concludes that

$$A(u - u_h, w_h) = 0, \text{ for all } w_h \in X_h. \tag{4.30}$$

Thus, we can write

$$
\begin{aligned}
\gamma \|u - u_h\|^2 &\leq A(u - u_h, u - u_h) \\
&= A(u - u_h, u - u_h) + A(u - u_h, u_h - v_h) \\
&= A(u - u_h, u - v_h) \leq \xi \|u - u_h\| \|u - v_h\|,
\end{aligned}
$$

where we used coercivity in the first line, the relation (4.30) with $w_h = u_h - v_h$ in the second line, and the linearity of A in its second argument and the continuity of A in the last line. Thus, we conclude that

$$
\|u - u_h\| \leq \frac{\xi}{\gamma} \|u - v_h\|, \text{ for all } v_h \in X_h,
$$

which implies that

$$
\|u - u_h\| \leq \frac{\xi}{\gamma} \inf_{v_h \in X_h} \|u - v_h\|. \tag{4.31}
$$

The term $\inf_{v_h \in X_h} \|u - v_h\|$ measures the "ability" of the finite element space X_h to approximate u. In particular, we say that the above family of finite element methods based on (4.23) has *the approximability property* if for any $v \in X$, we have

$$
\lim_{h \to 0} \inf_{v_h \in X_h} \|u - v_h\| = 0.
$$

The inequality (4.31) implies that *a family of discrete solutions associated to a family of conforming finite element methods for a coercive problem is convergent if that family has the approximability property.* This result is called *Céa's lemma.*

As mentioned earlier, the weak formulations of elliptic PDEs subject to the aforementioned boundary conditions are coercive problems with X being $H^1(\Omega)$ or one of its linear subspaces. Conforming finite element methods for these problem can then be obtained by using H^1-conformal finite element spaces X_h such as those associated to the n-simplex of type (k).

Let $I_h^k u$ be an interpolation of the solution u, where the interpolation operator I_h^k is associated to the n-simplex of type (k). Using the second inequality of (3.16), we conclude that if $\{T_h\}$ is a shape-regular family of meshes and if u is $s+1$ times differentialble with $1 \leq s \leq k$, we have

$$
\|u - u_h\|_{1,2} \leq \frac{\xi}{\gamma} \inf_{v_h \in X_h} \|u - v_h\|_{1,2} \leq \frac{\xi}{\gamma} \|u - I_h^k u\|_{1,2} \leq \tilde{C}(u) h^s, \tag{4.32}
$$

where $\tilde{C}(u)$ does not depend on h. Consequently, we have $\|u - u_h\|_{1,2} \to 0$ as $h \to 0$. *Notice that the convergence of finite element methods follows from the convergence of finite element interpolations.* This is the reason for studying finite element interpolations before finite element methods for PDEs.

Similar to finite element interpolations, the power of h in (4.32) is called *the convergence rate.* The maximum convergence rate of the n-simplex of type (k) in the H^1-norm is k, which occurs if u is $k+1$ times differentiable. It is also possible to show that

$$\|u - u_h\|_2 \leq C(u)h^{s+1}, \tag{4.33}$$

where $C(u)$ does not depend on h. Thus, the above finite element methods also converge in the L^2-norm. If u is $k+1$ times differentiable, one obtains the maximum convergence rate $k+1$ in the L^2-norm, which is called the *optimal* convergence rate.

4.6   IMPLEMENTATION: THE POISSON EQUATION

Now we discuss how to use FEniCS to solve an elliptic PDE subject to different boundary conditions. Consider the 2D Poisson's equation

$$-\Delta u = f \text{ in } \Omega, \tag{4.34}$$

where Ω is the unit square $(0,1) \times (0,1)$ and $f(x,y) = -4b^2(x^2+y^2)^{b-1}$, with b being a constant. A solution of this PDE is given by

$$u_e(x,y) = (x^2+y^2)^b. \tag{4.35}$$

Throughout this section, we solve (4.34) subject to different boundary conditions such that (4.35) is the solution of the associated boundary value problems.

To obtain a known *exact* solution such as (4.35) for the PDE (4.34), we begin by assuming any arbitrary form for the solution and then obtain the associated function f on the right side of (4.34) by simply plugging the assumed exact solution in the left side of (4.34). This way, one can always construct *a test problem* with a known solution for a given PDE. Test problems are very useful for debugging codes and for studying the performance of numerical methods since one can calculate the error of approximate solutions.

4.6.1   DIRICHLET BOUNDARY CONDITION

We begin by considering the 2D Poisson equation (4.34) subject to a Dirichlet boundary condition: Find u such that

$$\begin{cases} -\partial_{xx}u - \partial_{yy}u = f(x,y), & \text{in } \Omega, \\ u = u_e, & \text{on } \partial\Omega. \end{cases}$$

By using (4.11), the weak formulation of the above boundary value problem can be stated as: Find $u \in H^1(\Omega)$ satisfying $u = u_e$, on $\partial\Omega$, such that

$$\int_\Omega \nabla u \cdot \nabla v = \int_\Omega fv, \text{ for all } v \in H_0^1(\Omega), \tag{4.36}$$

where "·" on the left side denotes the inner product of vectors. Let V_h be the finite element space induced by a family of 2-simplices of type (k) on a mesh T_h of Ω. By using V_h and its subspace X_h defined in (4.24), we can define the following conformal finite element method based on (4.36): Find $u_h \in V_h$ with $u_h = I_h^k u_e$, on $\partial\Omega$, such that

$$\int_\Omega \nabla u_h \cdot \nabla v_h = \int_\Omega f v_h, \text{ for all } v_h \in X_h. \tag{4.37}$$

Consider a structured simplicial mesh for the unit square with N denoting the number of uniform divisions of each edge. Given N, the degree of finite element k, and the parameter b of the function f, the function Compute_Dirichlet computes the finite element solution of (4.37) and its L^2- and H^1-errors.

```python
def Compute_Dirichlet(N, k, b):
    """ Given mesh division size N, the degree k of Lagrange
    elements, and the parameter b of the exact solution, this
    function returns the mesh size h and errors of the finite
    element solution associated to a Dirichlet boundary
    condition """

    # create mesh and define function space
    mesh = UnitSquareMesh(N, N)
    V_h = FunctionSpace(mesh, "Lagrange", k)

    # the exact solution
    u_e = Expression('pow(x[0]*x[0] + x[1]*x[1], b)', b = b,
        degree = 3 + k)

    # define Dirichlet boundary condition
    def ue_boundary(x, on_boundary):
        return on_boundary

    BC = DirichletBC(V_h, u_e, ue_boundary)

    # define the weak formulation
    u_h = TrialFunction(V_h)
    v_h = TestFunction(V_h)
    f = Expression('-4*b*b*pow(x[0]*x[0] + x[1]*x[1], b-1)',
        b = b, degree = 3 + k)
    A = inner(grad(u_h), grad(v_h))*dx
    F = f*v_h*dx

    # compute finite element solution
    u_h = Function(V_h)    # u_h is redefined
    solve(A == F, u_h, BC)

    L2_Error = errornorm(u_e, u_h, norm_type="L2")
    H1_Error = errornorm(u_e, u_h, norm_type="H1")
```

```
return L2_Error, H1_Error, mesh.hmax()
```

This function works as follows: After defining the mesh mesh, the finite element space V_h, and the exact solution u_e, the Dirichlet boundary condition is defined by the following segment:

```
def ue_boundary(x, on_boundary):
    return on_boundary

BC = DirichletBC(V_h, u_e, ue_boundary)
```

The function ue_boundary is used to specify the region that the Dirichlet boundary condition will be applied. It returns the boolean value True if x is on the boundary and False otherwise. The argument on_boundary is provided by FEniCS and is True if x is on the boundary of a mesh. Since the Dirichlet boundary condition is applied on the whole boundary in the present problem, ue_boundary can simply return the value of on_boundary. Later, we will see less trivial examples of functions that only specify a specific portion of the boundary. The function DirichletBC defines the Dirichlet boundary condition using the value u_e and the function ue_boundary that specifies where the boundary condition should be applied.

From the mathematical point of view, the trial and the test spaces of the problem (4.37) are different and are obtained by subjecting V_h to 2 different boundary conditions. In FEniCS, boundary conditions are specified in the final stage of solving discrete equations and not in the definition of function spaces. Consequently, the definitions of the trial and the test functions are similar for the problem (4.37):

```
u_h = TrialFunction(V_h)
v_h = TestFunction(V_h)
```

After defining the function f, we define the left and the right sides of the problem (4.37):

```
A = inner(grad(u_h), grad(v_h))*dx
F = f*v_h*dx
```

The language employed in FEniCS to express weak forms is called UFL (Unified Form Language). UFL provides a very interesting feature of FEniCS: The above segment is very similar to the mathematical formula of the problem (4.37). The integration over domains is denoted by dx in UFL. As will be seen later, ds is used for the integration over domains' boundaries.

Having defined the problem (4.37), its solution is computed as:

```
u_h = Function(V_h)    # u_h is redefined
solve(A == F, u_h, BC)
```

In FEniCS, unknowns of weak formulations are `TrialFunction` objects and solutions are `Function` objects. In the mathematical statement (4.37), these concepts are the same and are denoted by u_h. After defining the bilinear form A, we do not need u_h as a `TrialFunction` object anymore. To be consistent with (4.37), we reused the variable name u_h and redefine it as a `Function` object to store the solution. Given the weak formulation A == F, the function object u_h, and the boundary condition BC, the function `solve` computes the finite element solution and stores it in u_h. Finally, the L^2- and H^1-errors are calculated by using `errornorm` and the results together with the mesh size `mesh.hmax()` are returned.

By default, `solve` uses sparse LU decomposition (Gaussian elimination) to solve linear systems. It is possible to change this option and use iterative methods such as preconditioned Krylov solvers, which are more appropriate for large problems.

One approach for debugging and checking finite element programs is to calculate errors and convergence rates associated to a family of meshes. For example, the following program computes the errors and convergence rates by using triangle of type (2), the parameter $b = 1.25$, and the meshes $N = 3, 6, 9, 12$.

```
from dolfin import *

# number of divisions of meshes
Divisions = [3, 6, 9, 12]

# computing errors
b = 1.25                        # parameter of the exact solution
k = 2                           # degree of Lagrange element
h = []                          # mesh sizes
error_L2, error_H1 = [], []     # initializing errors

for N in Divisions:
    Er_L2, Er_H1, hmesh = Compute_Dirichlet(N, k, b)
    error_L2.append(Er_L2); error_H1.append(Er_H1)
    h.append(hmesh)

# convergence rates
from math import log as ln  # log is a dolfin name too
rate_L2 = ln(error_L2[-1]/error_L2[-2])/ln(h[-1]/h[-2])
rate_H1 = ln(error_H1[-1]/error_H1[-2])/ln(h[-1]/h[-2])

# printing results
for i in range(len(Divisions)):
    print('N = %2.f, L2_error = %5.2E, H1_error = %5.2E'
          % (Divisions[i], error_L2[i], error_H1[i]))
print('L2-convergence rate = %.2f, H1-convergence rate = %.2f'
      % (rate_L2,rate_H1))
```

This program generates the following output:

```
N =  3, L2_error = 9.63E-04, H1_error = 1.94E-02
N =  6, L2_error = 1.21E-04, H1_error = 4.92E-03
N =  9, L2_error = 3.60E-05, H1_error = 2.20E-03
N = 12, L2_error = 1.52E-05, H1_error = 1.24E-03
L2-convergence rate = 3.00, H1-convergence rate = 1.99
```

The errors are decreasing and the computed convergence rates are consistent with the theoretical values 3 and 2 suggested by the estimates (4.33) and (4.32), respectively.

If the solution u belongs to the polynomial space of the underlying finite element, then the finite element solution u_h will be equal to u regardless of the refinement level of the underlying mesh. In this case, computed errors should be very small for any mesh, in the order of machine error. This is another approach for designing a test problem to verify finite element programs. For example, if we assume $b = 1$ in (4.35), then the exact solution will be a polynomial of degree 2. Therefore, the errors of finite element solutions induced by triangle of type (k) with $k \geq 2$ are expected to be very small. By setting b=1 in the above program, we get the following output, which is consistent with our expectation.

```
N =  3, L2_error = 1.38E-15, H1_error = 1.56E-14
N =  6, L2_error = 4.54E-15, H1_error = 3.59E-14
N =  9, L2_error = 9.11E-15, H1_error = 6.16E-14
N = 12, L2_error = 1.82E-14, H1_error = 1.05E-13
```

4.6.2 MIXED DIRICHLET-NEUMANN BOUNDARY CONDITION

Suppose Γ_D is the bottom edge of the unit square Ω and let $\Gamma_N = \Gamma_N^1 \cup \Gamma_N^2 \cup \Gamma_N^3$, where $\Gamma_N^1, \Gamma_N^2,$ and Γ_N^3 are respectively the right, the top, and the left edges of Ω. We solve (4.34) with the following mixed Dirichlet-Neumann boundary condition: Find u such that

$$
\begin{cases}
-\partial_{xx}u - \partial_{yy}u = f(x,y), & \text{in } \Omega, \\
u = u_e, & \text{on } \Gamma_D, \\
\partial_x u = \partial_x u_e, & \text{on } \Gamma_N^1 \text{ and } \Gamma_N^3, \\
\partial_y u = \partial_y u_e, & \text{on } \Gamma_N^2.
\end{cases}
\tag{4.38}
$$

By using (4.15), a weak formulation of the above problem can be stated as: Find $u \in H^1(\Omega)$ satisfying $u = u_e$, on Γ_D, such that

$$
\int_\Omega \nabla u \cdot \nabla v = \int_\Omega fv + \int_{\Gamma_N^1} v \partial_x u_e + \int_{\Gamma_N^2} v \partial_y u_e - \int_{\Gamma_N^3} v \partial_x u_e, \quad \text{for all } v \in H_D^1(\Omega).
$$

Similar to the previous section, let V_h be a finite element space induced by a family of 2-simplices of type (k) and consider the subspace X_h of V_h defined in (4.25). Then, we can write the following finite element method for the boundary value problem (4.38): Find $u_h \in V_h$ with $u_h = I_h^k u_e$, on Γ_D, such that

$$
\int_\Omega \nabla u_h \cdot \nabla v_h = \int_\Omega fv_h + \int_{\Gamma_N^1} v_h \partial_x u_e + \int_{\Gamma_N^2} v_h \partial_y u_e - \int_{\Gamma_N^3} v_h \partial_x u_e, \quad \text{for all } v_h \in X_h.
$$

The following Python function for solving this finite element method is the analog of Compute_Dirichlet defined in Section 4.6.1 for Dirichlet boundary conditions.

```python
def Compute_DN(N, k, b):
    """ Given mesh division size N, the degree k of Lagrange
    elements, and the parameter b of the exact solution, this
    function returns the mesh size h and errors of the finite
    element solution associated to a mixed Dirichlet-Nuemann
    boundary condtion """

    # create mesh and define function space
    mesh = UnitSquareMesh(N, N)
    V_h = FunctionSpace(mesh, "Lagrange", k)

    # the exact solution and its first derivatives
    u_e = Expression('pow(x[0]*x[0] + x[1]*x[1], b)', b = b,
                      degree = 3 + k)
    Dx_u_e = Expression("""2*b*x[0]*pow(x[0]*x[0]
                  + x[1]*x[1], b-1)""", b = b, degree = 3 + k)
    Dy_u_e = Expression("""2*b*x[1]*pow(x[0]*x[0]
                  + x[1]*x[1], b-1)""", b = b, degree = 3 + k)

    # marking the boundary using a mesh function
    boundary_parts = MeshFunction("size_t", mesh,
                                   mesh.topology().dim()-1)

    # mark right boundary edges as subdomain 1
    class RightEdges(SubDomain):
        def inside(self, x, on_boundary):
            tol = 1E-12    # tolerance for comparisons
            return on_boundary and abs(x[0] - 1) < tol
    Gamma_R = RightEdges()
    Gamma_R.mark(boundary_parts, 1)

    # mark top boundary edges as subdomain 2
    class TopEdges(SubDomain):
        def inside(self, x, on_boundary):
            tol = 1E-12    # tolerance for comparisons
            return on_boundary and abs(x[1] - 1) < tol
    Gamma_T = TopEdges()
    Gamma_T.mark(boundary_parts, 2)

    # mark left boundary edges as subdomain 3
    class LeftEdges(SubDomain):
        def inside(self, x, on_boundary):
```

```
        tol = 1E-12   # tolerance for comparisons
        return on_boundary and abs(x[0]) < tol
Gamma_L = LeftEdges()
Gamma_L.mark(boundary_parts, 3)

# define Dirichlet boundary condition
def bottom_boundary(x, on_boundary):
    tol = 1E-12   # tolerance for comparisons
    return on_boundary and abs(x[1]) < tol
BC = DirichletBC(V_h, u_e, bottom_boundary)

# define the weak formulation
u_h = TrialFunction(V_h)
v_h = TestFunction(V_h)
f = Expression('-4*b*b*pow(x[0]*x[0] + x[1]*x[1], b-1)',
               b = b, degree = 3 + k)
ds = Measure("ds", domain=mesh,
        subdomain_data=boundary_parts) # boundary integral
A = inner(grad(u_h), grad(v_h))*dx
F = f*v_h*dx + Dx_u_e*v_h*ds(1) + Dy_u_e*v_h*ds(2) \
    - Dx_u_e*v_h*ds(3)

# compute finite element solution
u_h = Function(V_h)         # u_h is redefined
solve(A == F, u_h, BC)

L2_Error = errornorm(u_e, u_h, norm_type="L2")
H1_Error = errornorm(u_e, u_h, norm_type="H1")

return L2_Error, H1_Error, mesh.hmax()
```

This function works as follows: After defining the mesh and the function spaces, the exact solution and its first derivatives are defined. The next step is to mark different parts of boundary for imposing the Nuemann boundary condition. This can be done by defining a function on boundary edges that takes the value i on Γ_N^i, $i = 1, 2, 3$. This function is defined by using the MeshFunction object:

```
boundary_parts = MeshFunction("size_t", mesh,
                        mesh.topology().dim()-1)
```

The argument mesh.topology().dim()-1 specifies that boundary_parts is defined on parts of the mesh of dimension 1 lower than the dimension of mesh. To assign values to boundary_parts, we use subclasses of the object SubDomain. For example, the following code assigns the value 1 to the right boundary Γ_N^1, where $x = 1$:

```
class RightEdges(SubDomain):
    def inside(self, x, on_boundary):
        tol = 1E-12    # tolerance for comparisons
        return on_boundary and abs(x[0] - 1) < tol
Gamma_R = RightEdges()
Gamma_R.mark(boundary_parts, 1)
```

The condition

```
on_boundary and abs(x[0] - 1) < tol
```

is True if x is on the boundary, that is on_boundary is True, and if x is "very close" to 1. Notice that using == to check the value of floating-point variables is not a good programming practice due to small round-off errors. Instead, we employed abs(x[0] - 1) < tol to check the equality $x = 1$ with a tolerance tol.

After marking Γ_N^1, Γ_N^2, and Γ_N^3, the Dirichlet boundary condition is defined on the bottom edge of the square where $y = 0$:

```
def bottom_boundary(x, on_boundary):
    tol = 1E-12    # tolerance for comparisons
    return on_boundary and abs(x[1]) < tol
BC = DirichletBC(V_h, u_e, bottom_boundary)
```

The next step is to define the weak formulation which involves integrals over the mesh as well as integrals over different parts of the mesh boundary. First, we use boundary_parts to define the symbol ds for the integration over the boundary:

```
ds = Measure("ds", domain=mesh,
             subdomain_data=boundary_parts) # boundary integral
```

Then, the symbols ds(1), ds(2), and ds(3) simply denote the integration over Γ_N^1, Γ_N^2, and Γ_N^3, respectively. These symbols are employed similar to dx to define the weak formulation:

```
A = inner(grad(u_h), grad(v_h))*dx
F = f*v_h*dx + Dx_u_e*v_h*ds(1) + Dy_u_e*v_h*ds(2) \
    - Dx_u_e*v_h*ds(3)
```

The rest of Compute_DN for solving the finite element method is similar to the function Compute_Dirichlet. By replacing Compute_Dirichlet with Compute_DN in the program of Section 4.6.1, we can compute errors and convergence rates for the mixed Dirichlet-Neumann boundary condition. For example, the output for $k = 2$, $b = 1.25$, and the meshes $N = 3, 6, 9, 12$, is:

```
N =  3, L2_error = 9.69E-04, H1_error = 1.84E-02
N =  6, L2_error = 1.21E-04, H1_error = 4.80E-03
N =  9, L2_error = 3.58E-05, H1_error = 2.16E-03
N = 12, L2_error = 1.51E-05, H1_error = 1.22E-03
L2-convergence rate = 3.00, H1-convergence rate = 1.98
```

We encourage the reader to compare this output with the corresponding output in Section 4.6.1.

4.6.3 ROBIN BOUNDARY CONDITION

Suppose $\partial\Omega = \Gamma^0 \cup \Gamma^1 \cup \Gamma^2 \cup \Gamma^3$, where Γ^0, Γ^1, Γ^2, and Γ^3 are respectively the bottom, the right, the top, and the left edges of the unit square Ω. Consider the following boundary value problem for (4.34) with a Robin boundary condition: Find u such that

$$\begin{cases} -\partial_{xx}u - \partial_{yy}u = f(x,y), & \text{in } \Omega, \\ u - \partial_y u = u_e - \partial_y u_e, & \text{on } \Gamma^0, \\ u + \partial_x u = u_e + \partial_x u_e, & \text{on } \Gamma^1, \\ u + \partial_y u = u_e + \partial_y u_e, & \text{on } \Gamma^2, \\ u - \partial_x u = u_e - \partial_x u_e, & \text{on } \Gamma^3. \end{cases} \tag{4.39}$$

By using (4.17), we can write the following weak formulation for this boundary value problem: Find $u \in H^1(\Omega)$ such that

$$\int_\Omega \nabla u \cdot \nabla v + \int_{\partial\Omega} uv = \int_\Omega fv + \int_{\Gamma^0}(u_e - \partial_y u_e)v + \int_{\Gamma^1}(u_e + \partial_x u_e)v$$
$$+ \int_{\Gamma^2}(u_e + \partial_y u_e)v + \int_{\Gamma^3}(u_e - \partial_x u_e)v, \text{ for all } v \in H^1(\Omega).$$

Let V_h be a finite element space induced by a family of 2-simplices of type (k). By using V_h, we can write the following finite element method for the boundary value problem (4.39): Find $u_h \in V_h$ such that

$$\int_\Omega \nabla u_h \cdot \nabla v_h + \int_{\partial\Omega} u_h v_h = \int_\Omega fv_h + \int_{\Gamma^0}(u_e - \partial_y u_e)v_h + \int_{\Gamma^1}(u_e + \partial_x u_e)v_h$$
$$+ \int_{\Gamma^2}(u_e + \partial_y u_e)v_h + \int_{\Gamma^3}(u_e - \partial_x u_e)v_h, \text{ for all } v_h \in V_h.$$

The Python function `Compute_Robin` for solving the above finite element method is obtained by some straightforward modifications of the Python function `Compute_DN` given in Section 4.6.2:

```
def Compute_Robin(N, k, b):
    """ Given mesh division size N, the degree k of Lagrange
    elements, and the parameter b of the exact solution, this
    function returns the mesh size h and errors of the FE
    solution associated to a Robin boundary condtion """

    # create mesh and define function space
    mesh = UnitSquareMesh(N, N)
    V_h = FunctionSpace(mesh, "Lagrange", k)

    # the exact solution and its first derivatives
    u_e = Expression('pow(x[0]*x[0] + x[1]*x[1], b)', b = b,
```

```
                     degree = 3 + k)
Dx_u_e = Expression("""2*b*x[0]*pow(x[0]*x[0]
             + x[1]*x[1], b-1)""", b = b, degree = 3 + k)
Dy_u_e = Expression("""2*b*x[1]*pow(x[0]*x[0]
             + x[1]*x[1], b-1)""", b = b, degree = 3 + k)

# marking the boundary using a mesh function
boundary_parts = MeshFunction("size_t", mesh,
                              mesh.topology().dim()-1)

# mark bottom boundary edges as subdomain 0
class BottomEdges(SubDomain):
    def inside(self, x, on_boundary):
        tol = 1E-12   # tolerance for comparisons
        return on_boundary and abs(x[1]) < tol
Gamma_B = BottomEdges()
Gamma_B.mark(boundary_parts, 0)

# mark right boundary edges as subdomain 1
class RightEdges(SubDomain):
    def inside(self, x, on_boundary):
        tol = 1E-12   # tolerance for comparisons
        return on_boundary and abs(x[0] - 1) < tol
Gamma_R = RightEdges()
Gamma_R.mark(boundary_parts, 1)

# mark top boundary edges as subdomain 2
class TopEdges(SubDomain):
    def inside(self, x, on_boundary):
        tol = 1E-12   # tolerance for comparisons
        return on_boundary and abs(x[1] - 1) < tol
Gamma_T = TopEdges()
Gamma_T.mark(boundary_parts, 2)

# mark left boundary edges as subdomain 3
class LeftEdges(SubDomain):
    def inside(self, x, on_boundary):
        tol = 1E-12   # tolerance for comparisons
        return on_boundary and abs(x[0]) < tol
Gamma_L = LeftEdges()
Gamma_L.mark(boundary_parts, 3)

# define the weak formulation
u_h = TrialFunction(V_h)
```

```
v_h = TestFunction(V_h)
f = Expression('-4*b*b*pow(x[0]*x[0] + x[1]*x[1], b-1)',
               b = b, degree = 3 + k)
ds = Measure("ds", domain=mesh,
        subdomain_data=boundary_parts) # boundary integral
A = inner(grad(u_h), grad(v_h))*dx + u_h*v_h*ds
F = f*v_h*dx + (u_e - Dy_u_e)*v_h*ds(0) \
    + (u_e + Dx_u_e)*v_h*ds(1) + (u_e + Dy_u_e)*v_h*ds(2) \
    + (u_e - Dx_u_e)*v_h*ds(3)

# compute finite element solution
u_h = Function(V_h)         # u_h is redefined
solve(A == F, u_h)

L2_Error = errornorm(u_e, u_h, norm_type="L2")
H1_Error = errornorm(u_e, u_h, norm_type="H1")

return L2_Error, H1_Error, mesh.hmax()
```

Notice that in the definition of the weak form, ds is used for the integration over the whole boundary while ds(0), ds(1), ds(2), and ds(3) denote the integration over specific portion of the boundary. Also notice that the solve command is simply solve(A == F, u_h), as there is no Dirichlet boundary condition. By replacing Compute_Dirichlet with Compute_Robin in the program of Section 4.6.1, we can compute errors and convergence rates for the Robin boundary condition. For example, the output for $k = 2$, $b = 1.25$, and the meshes $N = 3, 6, 9, 12$, is:

```
N =  3, L2_error = 8.64E-04, H1_error = 1.78E-02
N =  6, L2_error = 1.16E-04, H1_error = 4.71E-03
N =  9, L2_error = 3.56E-05, H1_error = 2.13E-03
N = 12, L2_error = 1.54E-05, H1_error = 1.21E-03
L2-convergence rate = 2.92, H1-convergence rate = 1.97
```

The reader is encouraged to compare this output with the corresponding outputs in Sections 4.6.1 and 4.6.2.

4.7 TIME-DEPENDENT PROBLEMS: PARABOLIC PROBLEMS

In this section, we study a class of time-dependent PDEs. Suppose Ω is an open domain in $\mathbb{R}^n$, $n = 1, 2, 3$, and let t indicate time in a time interval $[0, T]$. By using the notation for the elliptic PDE (4.1), we consider the time-dependent PDE

$$\partial_t u - \text{div}(\nabla u \cdot \mathbb{D}) + \mathbf{b} \cdot \nabla u + cu = f, \quad x \in \Omega, \, t \in [0, T], \qquad (4.40)$$

where $u : \Omega \times [0, T] \to \mathbb{R}$ is the time-dependent unknown function and the given data $f(x,t)$, $c(x,t)$, $\mathbf{b}(x,t)$, and $\mathbb{D}(x,t)$ are also time-dependent. We assume that for all

$\mathbf{w} \in \mathbb{R}^n$, there exists $\beta > 0$ such that

$$\mathbf{w} \cdot \mathbb{D}(x,t) \cdot \mathbf{w}^T \geq \beta \|\mathbf{w}\|^2, \text{ at any } x \in \Omega, \text{ and any } t \in [0,T]. \tag{4.41}$$

The PDE (4.40) is then called *parabolic*.

Example 4.3. The heat equation

$$\partial_t u - \text{div}(k(x)\nabla u) = f, \quad x \in \Omega, \ t \in [0,T], \tag{4.42}$$

is an important example of parabolic PDEs, where $u(x,t)$ is the temperature of the point x at time t, $k(x)$ is the thermal conductivity, and $f(x,t)$ is a source term.

To solve (4.42), one may specify the temperature $g(x,t)$ at the boundary $\partial\Omega$ for all t and also the initial distribution of temperature at time zero. These can be stated as

$$\begin{cases} u(x,t) = g(x,t), & x \in \partial\Omega, \ t \in [0,T], \\ u(x,0) = u_0(x), & x \in \Omega. \end{cases} \tag{4.43}$$

More generally, to solve (4.40), we can impose any of the boundary conditions introduced in Section 4.2 for $t \in [0,T]$ together with *the initial condition $u(x,0) = u_0(x)$, $x \in \Omega$*. As the following example demonstrates, weak formulations for time-dependent parabolic PDEs can be obtained by the approach of Section 4.2.

Example 4.4. Consider the PDE (4.40) subject to the boundary and the initial conditions (4.43) with $g(x,t) = 0$. By multiplying the PDE by an arbitrary $v \in H_0^1(\Omega)$ and using Green's formula, one obtains the following weak formulation: Find $u(x,t)$ such that $u(\cdot,t) \in H_0^1(\Omega)$ for any $t \in [0,T]$ and

$$\begin{cases} \int_\Omega v \partial_t u + B(t,u,v) = \int_\Omega f(x,t)v, \text{ for all } v \in H_0^1(\Omega), \\ u(x,0) = u_0(x), \end{cases}$$

where

$$B(t,u,v) = \int_\Omega \left\{ \nabla u \cdot \mathbb{D} \cdot (\nabla v)^T + (\boldsymbol{b} \cdot \nabla u)v + cuv \right\}.$$

Generally speaking, weak formulations of (4.40) subject to the boundary conditions of Section 4.2 and an initial condition at $t = 0$ can be stated by the abstract problem: Find $u(x,t)$ such that $u(\cdot,t) \in X$ for any $t \in [0,T]$ and

$$\begin{cases} \int_\Omega v \partial_t u + A(t,u,v) = F(t,v), \text{ for all } v \in X, \\ u(x,0) = u_0(x), \end{cases} \tag{4.44}$$

where $A(t,u,v)$ is linear with respect to u and v and $F(t,v)$ is linear with respect to v. The data X, A, and F are defined similar to those of Section 4.3 by using the time-dependent data $\mathbb{D}(x,t)$, $\boldsymbol{b}(x,t)$, $c(x,t)$, and $f(x,t)$. We leave the proof as an exercise.

4.7.1   FINITE ELEMENT APPROXIMATIONS USING THE METHOD OF LINES

Suppose T_h is a mesh of Ω and X_h is an H^1-conformal finite element space induced by T_h with $X_h \subset X$, where X is the trial and the test space of the abstract problem (4.44). We consider the following spatial discretization of (4.44): Find $u_h(x,t)$ such that $u_h(\cdot,t) \in X_h$ for any $t \in [0,T]$ and

$$\begin{cases} \int_\Omega v_h \partial_t u_h + A(t,u_h,v_h) = F(t,v_h), & \text{for all } v_h \in X_h, \\ u_h(x,0) = I_h u_0(x), \end{cases} \quad (4.45)$$

where $I_h u_0 \in X_h$ is the finite element interpolation of u_0.

The discrete problem (4.45) is equivalent to a linear system of ordinary differential equations (ODEs). To see this, notice that at any $t \in [0,T]$, we have $u_h(\cdot,t) \in X_h$, and therefore, we can write

$$u_h(x,t) = \sum_{j=1}^N U_j(t)\psi_j(x), \quad (4.46)$$

where $N = \dim X_h$, and $\{\psi_1,\ldots,\psi_N\}$ are the global shape functions of X_h. If u_h solves (4.45), then

$$\int_\Omega \psi_i \partial_t u_h + A(t,u_h,\psi_i) = F(t,\psi_i), \quad i=1,\ldots,N.$$

By using the expansion (4.46), the above relations can be stated as

$$\sum_{j=1}^N \partial_t U_j \int_\Omega \psi_j \psi_i + \sum_{j=1}^N U_j A(t,\psi_j,\psi_i) = F(t,\psi_i), \quad i=1,\ldots,N.$$

These equations form a system of coupled ODEs that admits the matrix form

$$\begin{cases} \mathbb{M}_{N\times N} \cdot \partial_t \mathbb{U}_{N\times 1}(t) = -[A]_{N\times N}(t) \cdot \mathbb{U}_{N\times 1}(t) + \mathbb{F}_{N\times 1}(t), & t \in [0,T], \\ \mathbb{U}(0) = \mathbb{U}_0, \end{cases} \quad (4.47)$$

where the symmetric matrix $\mathbb{M}_{N\times N}$ is *the mass matrix* with the components $\mathbb{M}_{ij} = \int_\Omega \psi_i \psi_j$, $[A]$ is the time-dependent stiffness matrix with the components $[A]_{ij} = A(t,\psi_j,\psi_i)$, and the time-dependent vectors $\mathbb{U}$ and $\mathbb{F}$ are given by

$$\mathbb{U}(t) = \begin{bmatrix} U_1(t) \\ \vdots \\ U_N(t) \end{bmatrix}, \ \mathbb{F}(t) = \begin{bmatrix} F(t,\psi_1) \\ \vdots \\ F(t,\psi_N) \end{bmatrix}.$$

The vector $\mathbb{U}_0$ contains the coefficients of the expansion of the initial value $I_h u_0$ in the basis $\{\psi_1,\ldots,\psi_N\}$.

By discretizing the system of ODEs (4.47) with respect to t, we can approximate its solution. In summary, to solve the problem (4.44), we first employed a spatial discretization to obtain (4.47) and then we can employ a temporal discretization to solve the discrete time-dependent problem (4.47). This approach is called *the method of lines*.

4.7.2 TEMPORAL DISCRETIZATION

Various finite-difference approaches are available for the temporal dicretization of the problem (4.45) or equivalently, the time-dependent ODEs (4.47). Here, we discuss two simple approaches, namely, *the explicit* and *the implicit Euler methods*.

Suppose $\{t_0, t_1, \ldots, t_m\}$ is a set of points with $t_0 = 0 < t_1 < t_2 < \cdots < t_m = T$, and let $\Delta_i t = t_{i+1} - t_i$. Also let $u_h^i(x) = u_h(x, t_i)$, that is, u_h^i is the value of u_h at *the time step t_i*. The explicit Euler method seeks the solution of (4.45) iteratively as follows: Let $u_h^0 = I_h u_0$. Given $u_h^i \in X_h$, find $u_h^{i+1} \in X_h$ such that

$$\frac{1}{\Delta_i t} \int_\Omega (u_h^{i+1} - u_h^i) v_h + A(t_i, u_h^i, v_h) = F(t_i, v_h), \text{ for all } v_h \in X_h. \qquad (4.48)$$

For the implicit Euler method, A and F are evaluated at t_{i+1}, that is: Let $u_h^0 = I_h u_0$. Given $u_h^i \in X_h$, find $u_h^{i+1} \in X_h$ such that

$$\frac{1}{\Delta_i t} \int_\Omega (u_h^{i+1} - u_h^i) v_h + A(t_{i+1}, u_h^{i+1}, v_h) = F(t_{i+1}, v_h), \text{ for all } v_h \in X_h. \quad (4.49)$$

It is possible to show that the explicit Euler method (4.48) and the implicit Euler method (4.49) converge to the solution of the abstract problem (4.44). Implicit Euler iterations involve solving a linear system and are computationally more expensive than explicit Euler iterations. However, one can use much larger time intervals $\Delta_i t$ using the implicit method which implies that fewer time steps are needed for the implicit Euler method. In particular, implicit methods are more suitable for stiff problems, where explicit methods require very fine time steps for accurate approximations.

4.7.3 IMPLEMENTATION: A DIFFUSION PROBLEM

We employ FEniCS to solve the following diffusion problem on the unit square $\Omega = (0,1) \times (0,1)$: Find $u(x,y,t)$ such that

$$\begin{cases} \partial_t u - \Delta u = f, & \text{in } \Omega, \ t \in [0,T], \\ u = u_e, & \text{on } \partial\Omega, \ t \in [0,T], \\ u = u_e, & \text{at } t = 0, \end{cases} \qquad (4.50)$$

with

$$f(x,y,t) = -4b^2(x^2 + y^2)^{b-1} + ct^{c-1},$$

where c and b are constants, and

$$u_e(x,y,t) = (x^2 + y^2)^b + t^c. \qquad (4.51)$$

It is straightforward to show that $u_e(x,y,t)$ is the solution of (4.50). A weak formulation for this problem reads: Find $u(x,y,t)$ satisfying $u(\cdot,t) \in H^1(\Omega)$ for any $t \in [0,T]$, with $u = u_e$, on $\partial\Omega$, such that

$$\begin{cases} \int_\Omega v \partial_t u + \int_\Omega \nabla u \cdot \nabla v = \int_\Omega f v, & \text{for all } v \in H_0^1(\Omega), \\ u = u_e, & \text{at } t = 0. \end{cases} \qquad (4.52)$$

Let V_h be the finite element space induced by 2-simplices of type (k) on a mesh T_h of Ω and let X_h be the subspace of V_h defined in (4.24). By using the implicit Euler method with a uniform time increment $S = t_{i+1} - t_i$, we can discretize (4.52) as: Let $u_h^0 = I_h^k u_e$ at $t = 0$. Given $u_h^i \in V_h$, find $u_h^{i+1} \in V_h$ satisfying $u_h^{i+1}|_{\partial\Omega} = I_h^k u_e$, such that

$$\int_\Omega u_h^{i+1} v_h + S \int_\Omega \nabla u_h^{i+1} \cdot \nabla v_h = S \int_\Omega f v_h + \int_\Omega u_h^i v_h, \text{ for all } v_h \in X_h. \quad (4.53)$$

The above framework is implemented in the following program.

```
""" This program solves a time-dependent diffusion problem.
    Given mesh division size N, the degree k of the Lagrange
    element, the parameters b and c of the exact solution,
    the final time T, and the time increment S, it computes
    the L2 error in each time step """

from dolfin import *
import numpy as np

# parameters
b = 1.0; c = 1.0        # for the exact solution
k = 2                   # degree of Lagrange element
N = 4                   # mesh refinement
T = 20                  # final time
S = 0.5                 # time increment
error = []              # initializing L2-errors

# create mesh and define function space
mesh = UnitSquareMesh(N, N)
V_h = FunctionSpace(mesh, "Lagrange", k)

# the exact solution
u_e = Expression('pow(x[0]*x[0] + x[1]*x[1], b) + pow(t, c)',
                 b = b, c = c, t = 0, degree = 3 + k)

# define Dirichlet boundary condition
def ue_boundary(x, on_boundary):
    return on_boundary
BC = DirichletBC(V_h, u_e, ue_boundary)

# initial condition
u0 = interpolate(u_e, V_h)

# define the weak formulation
u_h = TrialFunction(V_h)
v_h = TestFunction(V_h)
```

```
f = Expression("""-4*b*b*pow(x[0]*x[0] + x[1]*x[1], b-1) +
        c*pow(t, c-1)""", b = b, c = c, t = S, degree = 3 + k)
A = u_h*v_h*dx + S*inner(grad(u_h), grad(v_h))*dx
F = S*f*v_h*dx + u0*v_h*dx


# compute solution
u_h = Function(V_h)
t = S
while t <= T:
    # compute finite element solution
    u_e.t = t; f.t = t
    solve(A == F, u_h, BC)

    # compute error
    error.append(errornorm(u_e, u_h, norm_type="L2"))

    # updating for the next iteration
    t += S
    u0.assign(u_h)


# printing the maximum error
MaxError = np.array(error).max()
print("The maximum L2-error is %5.2E" % MaxError)
```

In this program, expressions, boundary conditions, and the weak formulation are defined as usual. Note that the time variable of the exact solution $u_e(x,y,t)$ and $f(x,y,t)$
are defined as a parameter of the corresponding Expression object:

```
u_e = Expression('pow(x[0]*x[0] + x[1]*x[1], b) + pow(t, c)',
                b = b, c = c, t = 0, degree = 3 + k)
f = Expression("""-4*b*b*pow(x[0]*x[0] + x[1]*x[1], b-1) +
        c*pow(t, c-1)""", b = b, c = c, t = S, degree = 3 + k)
```

The value of the time parameter can be modified by assigning values to u_e.t and
f.t. Time iterations are implemented by using the following while loop, where u0
and u_h respectively denote the solution of the previous and the current iterations.

```
u_h = Function(V_h)
t = S
while t <= T:
    # compute finite element solution
    u_e.t = t; f.t = t
    solve(A == F, u_h, BC)

    # compute error
    error.append(errornorm(u_e, u_h, norm_type="L2"))
```

```
# updating for the next iteration
t += S
u0.assign(u_h)
```

In each iteration, after updating values of the time parameters of u_e and f, the finite element solution u_h is calculated. At the end of the loop, the time variable t is updated to time of the next iteration and the current solution u_h is assigned to the solution of the previous time step u0. Here, t += S means t = t + S. After the loop, the maximum error of time steps is calculated by finding the maximum element of the list error and the result is printed as the output.

A test problem for the above program can be obtained by choosing the values $b = c = 1$ for the exact solution u_e and f. In this case, if we use Lagrange elements of degree 2, the exact solution will be also the solution of the discrete problem (4.53) and therefore, the errors should be very small, in the order of machine accuracy, regardless of the size of the mesh and the time increment S. To verify this, we run the program with the parameters $N = 4$, $S = 0.5$, and $T = 20$, which yields the expected output:

```
The maximum L2-error is 1.85E-14
```

By using a fixed time increment S, the stiffness matrix of the problem (4.53) solved in each iteration will be time-independent. Therefore, instead of assembling the stiffness matrix in each time step as we did in the above program, computationally it is more efficient to calculate the stiffness matrix once outside the loop and then use it in each iteration to compute solutions. To achieve this, we first place the following code before while:

```
Mat_A = assemble(A)    # assembling stiffness matrix only once
Vec_F = None           # for memory saving assemble
```

Then, instead of solve(A == F, u_h, BC), we use the following code inside the while loop:

```
Vec_F = assemble(F, tensor=Vec_F)
BC.apply(Mat_A, Vec_F)
solve(Mat_A, u_h.vector(), Vec_F)
```

Implementation of these changes in the above program is left as an exercise.

4.8 MIXED FINITE ELEMENT METHODS

Consider the following steady-state heat equation with a Dirichlet boundary condition: Find the temperature distribution u of a body Ω such that $u = g$ on $\partial\Omega$ and

$$-\text{div}\left(\kappa \nabla u\right) = f \text{ in } \Omega, \tag{4.54}$$

where $\kappa(x)$ is the thermal conductivity, $f(x)$ is a known heat source, and $g(x)$ is a known temperature distribution at the boundary. Suppose that in addition to the temperature distribution u, we are also interested to find the heat flow

$$q = -\kappa \nabla u. \qquad (4.55)$$

To this end, we can employ a finite element method to obtain a finite element approximation u_h of u and then approximate q by using u_h. A simple approach is to simply set $\bar{q}_h = -\kappa \nabla u_h$, that is, the value of $\bar{q}_h$ on each element is calculated by using the gradient of u_h on that element. Another approach is to solve the following weak formulation of (4.55): Given u_h, find q such that

$$\int_\Omega q \cdot p = -\int_\Omega \kappa \nabla u_h \cdot p, \text{ for all vector fields } p. \qquad (4.56)$$

Let $\hat{q}_h$ be the finite element solution of (4.56). The solution $\hat{q}_h$ is the *projection* of $-\kappa \nabla u_h$ on the associated finite element space. Generally speaking, the approximations $\bar{q}_h$ and $\hat{q}_h$ are not equal and both may be a poor approximation of q. As will be discussed in the sequel, an alternative approach for approximating q is to use *a mixed formulation* for the problem (4.54) which approximates u and q simultaneously.

4.8.1   MIXED FORMULATIONS

To approximate the heat flow q, we may consider it as a separate unknown and calculate u and q simultaneously. To this end, instead of (4.54), we consider the following problem: Find the temperature distribution u and the heat flow q such that $u = g$ on $\partial\Omega$ and

$$\left.\begin{array}{l} \operatorname{div} q = f, \\ q + \kappa \nabla u = 0, \end{array}\right\} \text{ in } \Omega. \qquad (4.57)$$

To write a weak formulation for the equations (4.57), we notice that we can assume $u \in H^1(\Omega)$ and $q \in H(\operatorname{div};\Omega)$, where the latter space was defined in Example 2.9. By multiplying the first equation of (4.57) by an arbitrary test function $v \in H_0^1(\Omega)$, taking the integral over Ω, and applying Green's formula, we can write

$$\int_\Omega v \operatorname{div} q = -\int_\Omega q \cdot \nabla v = \int_\Omega f v.$$

Similarly, we multiply the second equation of (4.57) by an arbitrary test vector field $p \in H(\operatorname{div};\Omega)$ and take the integral over Ω. Consequently, we obtain the following weak formulation of (4.57): Find $u \in H^1(\Omega)$ and $q \in H(\operatorname{div};\Omega)$ such that $u = g$ on $\partial\Omega$ and

$$\begin{aligned} \int_\Omega q \cdot \nabla v &= -\int_\Omega f v, &\text{for all } v \in H_0^1(\Omega), \\ \int_\Omega q \cdot p + \int_\Omega \kappa \nabla u \cdot p &= 0, &\text{for all } p \in H(\operatorname{div};\Omega). \end{aligned} \qquad (4.58)$$

The weak formulation (4.58) that involves simultaneous approximations of an unknown and its derivatives is called *a mixed formulation* for (4.54). In the literature, the term mixed formulation is usually used to refer to weak formulations involving two or more unknowns that have a *saddle-point* variational structure such as Stokes equation in fluid mechanics in terms of velocity and pressure. Here, solutions are a saddle-point of an associated functional. We will consider this latter case in Section 5.8.

4.8.2   MIXED METHODS AND INF-SUP CONDITIONS

Let T_h be a simplicial mesh of Ω. Suppose V_h is the finite element space induced by 2-simplex of type (k) and let X_h be the subspace of V_h defined in (4.24). Also suppose W_h is the finite element space induced by the 2D Raviart-Thomas finite element. A conformal finite element method for the mixed formulation (4.58) reads: Find $u_h \in V_h$ and $\boldsymbol{q}_h \in W_h$ such that $u_h = I_h^k g$ on $\partial\Omega$ and

$$
\begin{aligned}
\int_\Omega \boldsymbol{q}_h \cdot \nabla v_h &= -\int_\Omega f v_h, &&\text{for all } v \in X_h, \\
\int_\Omega \boldsymbol{q}_h \cdot \boldsymbol{p}_h + \int_\Omega \kappa \nabla u_h \cdot \boldsymbol{p}_h &= 0, &&\text{for all } \boldsymbol{p}_h \in W_h.
\end{aligned}
\tag{4.59}
$$

This finite element method is called *a mixed finite element method* for the problem (4.54) as it is based on a mixed formulation.

An important difference between problems based on mixed formulations and the coercive problems discussed earlier is that *discretizations of a well-posed mixed problem may not be well-posed*. For the mixed finite element method (4.59), this fact can be explained by using the associated stiffness matrix. In particular, let $\{\psi_1, \ldots, \psi_N\}$ and $\{\boldsymbol{\phi}_1 \ldots, \boldsymbol{\phi}_M\}$ be respectively the global shape functions of X_h and W_h with $\dim X_h = N$, and $\dim W_h = M$. Using

$$
u_h = \sum_{j=1}^N U_j \psi_j, \text{ and } \boldsymbol{q}_h = \sum_{j=1}^N Q_j \boldsymbol{\phi}_j,
$$

it is easy to show that solving (4.59) leads to a linear system

$$
\mathbf{S}_{K \times K} \cdot \mathbb{V}_{K \times 1} = \mathbb{F}_{K \times 1},
\tag{4.60}
$$

where $K = N + M$, and $\mathbf{S}$ and $\mathbb{V}$ are respectively the stiffness matrix and the unknown vector. The stiffness matrix $\mathbf{S}$ and the vectors $\mathbb{V}$ and $\mathbb{F}$ are of the forms

$$
\mathbf{S} = \begin{bmatrix} \mathbf{0} & \mathbf{S}_{N \times M}^{1d} \\ \hline \mathbf{S}_{M \times N}^{d1} & \mathbf{S}_{M \times M}^{dd} \end{bmatrix}, \quad \mathbb{V} = \begin{bmatrix} \mathbb{U}_{N \times 1} \\ \mathbb{Q}_{M \times 1} \end{bmatrix}, \quad \mathbb{F} = \begin{bmatrix} \hat{\mathbb{F}}_{N \times 1} \\ \bar{\mathbb{F}}_{M \times 1} \end{bmatrix},
\tag{4.61}
$$

where $\mathbb{U} = [U_1, \ldots, U_N]^T$, $\mathbb{Q} = [Q_1, \ldots, Q_M]^T$, and the components of the submatrices $\mathbf{S}^{1d}$, $\mathbf{S}^{d1}$, and $\mathbf{S}^{dd}$ of the stiffness matrix are given by

$$
\mathbf{S}_{ij}^{1d} = \int_\Omega \boldsymbol{\phi}_j \cdot \nabla \psi_i, \quad \mathbf{S}_{ij}^{d1} = \int_\Omega \kappa \boldsymbol{\phi}_i \cdot \nabla \psi_j, \quad \mathbf{S}_{ij}^{dd} = \int_\Omega \boldsymbol{\phi}_i \cdot \boldsymbol{\phi}_j.
$$

If the mixed finite element method (4.59) is well-posed, then the stiffness matrix $\mathbf{S}$ is non-singular and (4.60) has a unique solution $\mathbb{V}$ for any $\mathbb{F}$. Then, by expanding the left side of (4.60) and using (4.61), we conclude that for any $\hat{\mathbb{F}} \in \mathbb{R}^N$ there exists $\mathbb{Q} \in \mathbb{R}^M$ such that

$$\mathbf{S}^{1d} \cdot \mathbb{Q} = \hat{\mathbb{F}},$$

since (4.60) has a solution for any $\mathbb{F}$. Hence, the matrix $\mathbf{S}^{1d}$ should be full rank, or equivalently, if we consider the matrix $\mathbf{S}^{1d}$ as a linear mapping $\mathbb{R}^M \to \mathbb{R}^N$, then $\mathbf{S}^{1d}$ should be surjective. Due to the rank-nulity theorem, we then conclude that $M \geq N$, see Exercises 2.13 and 2.14. *In summary, we showed that the mixed finite element method* (4.59) *does not admit a unique solution if* $\dim W_h < \dim X_h$.

Example 4.5. The choice of 2-simplex of type (2) and the Raviart-Thomas element (of degree 1) in (4.59) leads to an unstable finite element method since $\dim W_h < \dim X_h$. To see this, let n_v^∂ denote the number of vertices on the boundary. The discussion of Example 3.7 implies that $\dim X_h = n_v + n_e - n_v^\partial$. On the other hand, from Section 3.4.2 we know that $\dim W_h = n_e$. Since $n_v - n_v^\partial > 0$ for practically useful meshes, we conclude that $\dim W_h < \dim X_h$ for such meshes.

The condition $M \geq N$ is only a necessary condition for the surjectivity of the linear mapping $\mathbf{S}^{1d} : \mathbb{R}^M \to \mathbb{R}^N$. *Necessary and sufficient* conditions for the surjectivity of linear mappings can be expressed by using *inf-sup conditions*. In particular, $\mathbf{S}^{1d}$ is surjective if and only if there exists $\alpha > 0$ such that

$$\inf_{v_h \in X_h} \sup_{\boldsymbol{q}_h \in W_h} \frac{\int_\Omega \boldsymbol{q}_h \cdot \nabla v_h}{\|\boldsymbol{q}_h\|_d \, \|v_h\|_{1,2}} \geq \alpha, \tag{4.62}$$

where sup and inf were defined in Section 2.1 and the norms $\|v_h\|_{1,2}$ and $\|\boldsymbol{q}_h\|_d$ are respectively defined in Examples 2.27 and 2.28. The left side of the above condition is interpreted as follows: For any *non-zero* members $v_h \in X_h$ and $\boldsymbol{q}_h \in W_h$, consider the real-valued function

$$r(v_h, \boldsymbol{q}_h) = \frac{\int_\Omega \boldsymbol{q}_h \cdot \nabla v_h}{\|\boldsymbol{q}_h\|_d \, \|v_h\|_{1,2}}.$$

To evaluate the left side of (4.62), we first obtain the the the set $E(v_h)$ for any non-zero v_h by calculating the supremum $E(v_h) = \sup\{r(v_h, \boldsymbol{q}_h) | \boldsymbol{q}_h \in W_h, \, \boldsymbol{q}_h \neq 0\}$, and then we obtain the infimum $\inf\{E(v_h) | v_h \in X_h, v_h \neq 0\}$. It is possible to use singular-values to computationally verify the inf-sup condition (4.62). We will not discuss this topic in more detail and refer the reader to references mentioned at the end of this chapter.

4.8.3   IMPLEMENTATION

It is easy to implement mixed finite element methods in FEniCS. As an example, we implement a mixed finite element method for the 2D Poisson equation (4.34), that is, we solve (4.57) on the unit square $(0,1) \times (0,1)$ where $\kappa(x) = 1$, and $f(x,y) = -4b^2(x^2+y^2)^{b-1}$, with b being a constant. The exact solution of this problem is

$$u_e(x,y) = (x^2+y^2)^b, \quad \boldsymbol{q}_e(x,y) = -2b(x^2+y^2)^{b-1} \begin{bmatrix} x \\ y \end{bmatrix}.$$

The mixed finite element method (4.59) then reads: Find $u_h \in V_h$ and $\boldsymbol{q}_h \in W_h$ such that $u_h = I_h^k u_e$, on $\partial\Omega$ and

$$\int_\Omega \boldsymbol{q}_h \cdot \nabla v_h = -\int_\Omega f v_h, \qquad \text{for all } v \in X_h,$$

$$\int_\Omega \boldsymbol{q}_h \cdot \boldsymbol{p}_h + \int_\Omega \nabla u_h \cdot \boldsymbol{p}_h = 0, \quad \text{for all } \boldsymbol{p}_h \in W_h.$$

$$(4.63)$$

This mixed finite element method is implemented in the following function.

```
def Compute_Mixed(N, k, b):
    """ Given mesh division size N, the degree k of Lagrange
    elements, and the parameter b of the exact solution, this
    function returns the mesh size h and errors associated to
    a mixed finite element method for the Poisson equation
    """

    # create mesh and define function space
    mesh = UnitSquareMesh(N, N)
    LG = FiniteElement("Lagrange", mesh.ufl_cell(), k)
    RT = FiniteElement("RT", mesh.ufl_cell(), 1)
    Z_h = FunctionSpace(mesh, MixedElement([LG, RT]))

    # the exact solution
    u_e = Expression('pow(x[0]*x[0] + x[1]*x[1], b)', b = b,
                     degree = 3 + k)
    q_e = Expression(
        ('-2*b*pow(x[0]*x[0] + x[1]*x[1], b-1)*x[0]',
         '-2*b*pow(x[0]*x[0] + x[1]*x[1], b-1)*x[1]'), b = b,
        degree = 4)

    # define Dirichlet boundary condition
    def ue_boundary(x, on_boundary):
        return on_boundary
    BC = DirichletBC(Z_h.sub(0), u_e, ue_boundary)

    # define the weak formulation
    (u_h, q_h) = TrialFunctions(Z_h)     # trial functions
    (v_h, p_h) = TestFunctions(Z_h)      # test fucntions
    f = Expression('-4*b*b*pow(x[0]*x[0] + x[1]*x[1], b-1)',
```

```
                        b = b, degree = 3 + k)
    LH = ( inner(q_h, grad(v_h)) + inner(q_h, p_h) +
           inner(grad(u_h), p_h) )*dx    # left side
    RH = -f*v_h*dx                       # right side


    # compute finite element solution
    z_h = Function(Z_h)
    solve(LH == RH, z_h, BC)
    (u_h, q_h) = z_h.split()


    Error_u = errornorm(u_e, u_h, norm_type="L2")
    Error_q = errornorm(q_e, q_h, norm_type="L2")

    return Error_u, Error_q, mesh.hmax()
```

The algorithm of this function is similar to that of the function Compute_Dirichlet of Section 4.6.1. The finite element space for this mixed method is defined by using MixedElement:

```
Z_h = FunctionSpace(mesh, MixedElement([LG, RT]))
```

This approach can be used for mixed problems with more than two unknowns as well. Alternatively, we can also define Z_h as:

```
Z_h = FunctionSpace(mesh, LG*RT)
```

However, the latter approach is not easily extended to mixed problems with more than two unknowns. To access the first or the second parts of Z_h, one can use Z_h.sub(i) where i is 0 or 1. The Dirichlet boundary condition is defined on the first part of Z_h by:

```
BC = DirichletBC(Z_h.sub(0), u_e, ue_boundary)
```

Trial and test functions associated to different parts of Z_h are defined by using TrialFunctions and TestFunctions:

```
(u_h, q_h) = TrialFunctions(Z_h)    # trial functions
(v_h, p_h) = TestFunctions(Z_h)     # test fucntions
```

Notice that for standard problems with one unknown, we use TrialFunction and TestFunction.

Weak formulations are defined by a single equation in FEniCS. For the mixed formulation (4.63), this can be achieved by noticing that the problem (4.63) can be expressed by a single equation as follows: Find $u_h \in V_h$ and $q_h \in W_h$ such that $u_h = I_h^k u_e$, on $\partial\Omega$ and

$$\int_\Omega q_h \cdot \nabla v_h + \int_\Omega q_h \cdot p_h + \int_\Omega \nabla u_h \cdot p_h = -\int_\Omega f v_h, \quad \text{for all } v \in X_h, \, p_h \in W_h.$$

This equation is obtained by adding the two equations of (4.63) and is implemented as:

```
f = Expression('-4*b*b*pow(x[0]*x[0] + x[1]*x[1], b-1)',
               b = b, degree = 3 + k)
LH = ( inner(q_h, grad(v_h)) + inner(q_h, p_h) +
       inner(grad(u_h), p_h) )*dx    # left side
RH = -f*v_h*dx                       # right side
```

The following code is employed to solve the mixed problem and obtain the solutions u_h and $\boldsymbol{q}_h$:

```
z_h = Function(Z_h)
solve(LH == RH, z_h, BC)
(u_h, q_h) = z_h.split()
```

The following program uses `Compute_Mixed(N, k, b)` to compute errors and convergence rates associated to a family of meshes.

```
# number of divisions of meshes
Divisions = [3, 6, 9, 12]

# computing errors
b = 1.25                     # parameter of the exact solution
k = 1                        # degree of Lagrange element
h = []                       # mesh sizes
error_u, error_q = [], []    # initializing errors

for N in Divisions:
    Er_u, Er_q, hmesh = Compute_Mixed(N, k, b)
    error_u.append(Er_u); error_q.append(Er_q)
    h.append(hmesh)

# convergence rates
from math import log as ln  # log is a dolfin name too
rate_u = ln(error_u[-1]/error_u[-2])/ln(h[-1]/h[-2])
rate_q = ln(error_q[-1]/error_q[-2])/ln(h[-1]/h[-2])

# printing results
for i in range(len(Divisions)):
    print('N = %2.f, error_u = %5.2E, error_q = %5.2E'
          % (Divisions[i], error_u[i], error_q[i]))
print('rate_u = %.2f, rate_q = %.2f' % (rate_u,rate_q))
```

The output of the above program with $b = 1.25$, $k = 1$, and a family of meshes with number of edge divisions $N = 3, 6, 9, 12$, reads:

```
N =  3, error_u = 4.34E-02, error_q = 2.02E-01
N =  6, error_u = 9.76E-03, error_q = 7.42E-02
N =  9, error_u = 4.14E-03, error_q = 4.31E-02
N = 12, error_u = 2.26E-03, error_q = 2.99E-02
rate_u = 2.10, rate_q = 1.27
```

As mentioned earlier, we can also use a single-field finite element method for u and then approximate ∇u by using the approximation of u. In the following function, we first solve a single-field finite element method for (4.54) by using the data of this section and then we calculate $q = -\nabla u$ by using the projection (4.56).

```python
def Compute_Grad(N, k, b):
    """ Given mesh division size N, the degree k of Lagrange
    elements, and the parameter b of the exact solution, this
    function returns the mesh size h, the error of the finite
    element solution u_h, and the error of -grad(u_h) """

    # create mesh and define function space
    mesh = UnitSquareMesh(N, N)
    V_h = FunctionSpace(mesh, "Lagrange", k)
    Vec_h = FunctionSpace(mesh, "RT", 1)   # FE space for grad

    # the exact solution
    u_e = Expression('pow(x[0]*x[0] + x[1]*x[1], b)', b = b,
                    degree = 3 + k)
    q_e = Expression(
        ('-2*b*pow(x[0]*x[0] + x[1]*x[1], b-1)*x[0]',
        '-2*b*pow(x[0]*x[0] + x[1]*x[1], b-1)*x[1]'), b = b,
        degree = 4)

    # define Dirichlet boundary condition
    def ue_boundary(x, on_boundary):
        return on_boundary
    BC = DirichletBC(V_h, u_e, ue_boundary)

    # define the weak formulation
    u_h = TrialFunction(V_h)
    v_h = TestFunction(V_h)
    f = Expression('-4*b*b*pow(x[0]*x[0] + x[1]*x[1], b-1)',
                    b = b, degree = 3 + k)
    A = inner(grad(u_h), grad(v_h))*dx
    F = f*v_h*dx

    # compute finite element solution
    u_h = Function(V_h)        # u_h represents the solution
```

```
solve(A == F, u_h, BC)

q_h = project(-grad(u_h), Vec_h) # q using projection

Error_u = errornorm(u_e, u_h, norm_type="L2")
Error_q = errornorm(q_e, q_h, norm_type="L2")

return Error_u, Error_q, mesh.hmax()
```

This function works similar to the function `Compute_Dirichlet` of Section 4.6.1. After obtaining the finite element solution u_h, the gradient q_h is obtained by projecting -grad(u_h) on the Raviart-Thomas finite element space Vec_h by using the FEniCS function `project`:

```
q_h = project(-grad(u_h), Vec_h) # q using projection
```

If we use `Compute_Grad` instead of `Compute_Mixed` to calculate errors and convergence rates associated to $b = 1.25$, $k = 1$, and a family of meshes with number of edge divisions $N = 3, 6, 9, 12$, we will obtain:

```
N =  3, error_u = 5.76E-02, error_q = 2.27E-01
N =  6, error_u = 1.44E-02, error_q = 8.27E-02
N =  9, error_u = 6.43E-03, error_q = 4.74E-02
N = 12, error_u = 3.62E-03, error_q = 3.25E-02
rate_u = 2.00, rate_q = 1.31
```

By comparing the above results with those obtained using `Compute_Mixed`, we observe that the mixed finite element method reduces the approximation errors of *both* u and q.

EXERCISES

Exercise 4.1. Suppose the matrix $\mathbb{D}$ in (4.1) is symmetric and thus, all its eigenvalues are real numbers. Show that the ellipticity condition (4.2) is then equivalent to the positive-definiteness of $\mathbb{D}$ (that is, all eigenvalues are positive) with the smallest eigenvalue being greater than or equal to β.

Exercise 4.2. Show that B defined in (4.8) is a bilinear form.

Exercise 4.3. Derive the weak formulation (4.10) for non-homogeneous Dirichlet boundary conditions.

Exercise 4.4. Derive the weak formulation (4.13) for Neumann boundary conditions.

Exercise 4.5. Derive the weak formulation (4.15) for the mixed Dirichlet-Neumann boundary condition (4.14).

Exercise 4.6. Derive the weak formulation (4.17) for the Robin boundary condition (4.16).

Exercise 4.7. Derive the relation (4.22).

Exercise 4.8. Show that if the bilinear form A is symmetric and positive, then a minimizer of the functional (4.21) is a solution of the problem (4.18).

Exercise 4.9. Specify sufficient conditions that guarantee the weak formulations of elliptic PDEs introduced in Section 4.2 are also variational formulations. For each boundary condition, determine the functional of the associated minimization problem.

Exercise 4.10. Suppose the bilinear form $A : X_h \times X_h \to \mathbb{R}$ of the discrete problem (4.23) is symmetric and satisfies the coercivity condition (4.28). Show that $[A]$ is positive definite in the sense of (4.29). Also show that all eigenvalues of $[A]$ are greater than or equal to γ, where γ is the coercivity constant in (4.28).

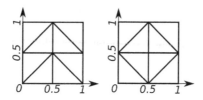

Figure 4.1 Meshes of the unit square.

Exercise 4.11. Consider the uniform meshes of the unit square $(0,1) \times (0,1)$ shown in Figure 4.1. Using 2-simplex of type (1) and these meshes, find the expression of the associated finite element solutions of the boundary value problem with the weak form (4.36) for the 2D Poisson equation, where the parameter b of u_e in (4.35) is assumed to be 1.25. Which mesh yields more accurate approximation of the value of u at the center point $\left(\frac{1}{2}, \frac{1}{2}\right)$?

Exercise 4.12. Repeat Exercise 4.11 for the boundary value problem (4.38).

Exercise 4.13. Repeat Exercise 4.11 for the boundary value problem (4.39).

Exercise 4.14. Consider the Poisson equation $-\Delta u = f$, over the unit cube $\Omega = (0,1) \times (0,1) \times (0,1)$. Determine the function f such that $u_e(x,y,z) = (x^2 + y^2 + z^2)^b$, is a solution, where b is a constant.

Exercise 4.15. Repeat Exercise 4.14 for $u_e(x,y,z) = \sin x \, \sin y \, \sin z$.

Exercise 4.16. Verify that weak formulations of (4.40) subject to the boundary conditions of Section 4.2 and an initial condition can be stated as the abstract problem (4.44). Write the data X, A, and F for each boundary condition.

Exercise 4.17. A *hyperbolic* PDE is a PDE of the form

$$\partial_{tt}u - \text{div}(\nabla u \cdot \mathbb{D}) + \boldsymbol{b} \cdot \nabla u + cu = f, \quad x \in \Omega, \, t \in [0,T], \qquad (4.64)$$

where we used the notation of (4.40) and $\mathbb{D}$ satisfies (4.41). To obtain an initial-boundary value problem for (4.64), we impose any of the boundary conditions of Section 4.2 for $t \in [0,T]$ together with *the initial conditions* $u(x,0) = u_0(x)$, and $\partial_t u(x,0) = w_0(x)$, $x \in \Omega$, where u_0 and w_0 are given. Show that weak formulations of this boundary value problem can be stated as the abstract problem

$$\begin{cases} \int_\Omega v \partial_{tt}u + A(t,u,v) = F(t,v), \text{ for all } v \in X, \\ u(x,0) = u_0(x), \quad \partial_t u(x,0) = w_0(x), \end{cases} \qquad (4.65)$$

where $A(t,u,v)$ is linear with respect to u and v and $F(t,v)$ is linear with respect to v. Write the data X, A, and F for each boundary condition of Section 4.2.

Exercise 4.18. By applying the method of lines to (4.64) obtain the analog of the ODEs (4.47) for hyperbolic PDEs.

Exercise 4.19. Consider times steps $t_0 = 0 < t_1 < \cdots < t_m = T$, with the uniform time step $S = t_{i+1} - t_i$. Using the three-point centered-difference formula

$$\partial_{tt}u(x,t_i) \approx \frac{u(x,t_i - S) - 2u(x,t_i) + u(x,t_i + S)}{S^2},$$

obtain implicit and explicit finite element methods for approximating a solution of the weak form (4.65) of hyperbolic PDEs.

Exercise 4.20. By using an argument similar to that of Example 4.5 show that the choice of 2-simplex of type (3) and the Raviart-Thomas element (of degree 1) in (4.59) leads to unstable finite element methods.

COMPUTER EXERCISES

Computer Exercise 4.1. By extending the programs of Section 4.6.1, solve the 3D Poisson equation of Exercise 4.14 subject to the Dirichlet boundary condition $u = u_e$, on $\partial\Omega$ by using 3-simplex of type (1). Use the division numbers $N = 2,3,4$, and $b = 1.25$, to compute L^2- and H^1-errors and the associated convergence rates.

Computer Exercise 4.2. Repeat Computer Exercise 4.1 by using $b = 0.75$, and $b = 0.25$. Are the convergence rates similar to those of Computer Exercise 4.1? Why?

Computer Exercise 4.3. Repeat Computer Exercise 4.1 by using 3-simplex of type (2).

Computer Exercise 4.4. Repeat Computer Exercise 4.1 by using 3-simplex of type (2) and $b = 1$. Are the calculated convergence rates meaningful? Justify your answer.

Computer Exercise 4.5. By extending the program of Section 4.6.2 solve the 3D Poisson equation of Exercise 4.14 subject to a mixed Dirichlet-Neumann boundary condition similar to that of (4.38), where the Dirichlet boundary condition $u = u_e$, is imposed on the bottom boundary $z = 0$, and the Nuemann boundary condition for normal derivatives is imposed on the remaining boundary faces. Use 3-simplex of type (1), the division numbers $N = 2, 3, 4$, and $b = 1.25$, to compute L^2- and H^1-errors and the associated convergence rates.

Computer Exercise 4.6. By extending the program of Section 4.6.3 solve the 3D Poisson equation of Exercise 4.14 subject to a Robin boundary condition similar to that of (4.39). Use 3-simplex of type (1), the division numbers $N = 2, 3, 4$, and $b = 1.25$, to compute L^2- and H^1-errors and the associated convergence rates.

Computer Exercise 4.7. By modifying the program of Section 4.7.3, solve the Diffusion problem (4.50) using the explicit Euler method.

Computer Exercise 4.8. Extend the program of Section 4.7.3 to solve the Diffusion problem (4.50) on the unit cube $\Omega = (0, 1) \times (0, 1) \times (0, 1)$, where the function $f(x, y, z, t)$ corresponds to the exact solution $u_e(x, y, z, t) = (x^2 + y^2 + z^2)^b + t^c$. Verify your program by solving a test problem.

Computer Exercise 4.9. Following the discussion at the end of Section 4.7.3, modify the program of that section such that the time-independent stiffness matrix is assembled only once outside time iterations.

Computer Exercise 4.10. *The wave equation* is an example of hyperbolic PDEs introduced in Exercise 4.17. Consider the following initial-boundary value problem for the wave equation on the unit square $\Omega = (0, 1) \times (0, 1)$: Find $u(x, y, t)$ such that

$$
\begin{cases}
\partial_{tt} u - \Delta u = f, & \text{in } \Omega, \ t \in [0, T], \\
u = u_e, & \text{on } \partial\Omega, \ t \in [0, T], \\
u = u_e, & \text{at } t = 0, \\
\partial_t u = \partial_t u_e, & \text{at } t = 0,
\end{cases}
\tag{4.66}
$$

where u_e is given in (4.51) and $f(x, y, t)$ is defined such that u_e is the solution of (4.66). By using the implicit finite element method of Exercise 4.19 and FEniCS solve (4.66). Assume $b = 1$, $c = 2$, and use 2-simplex of type (2).

Computer Exercise 4.11. Develop a mixed finite element method for the 3D Poisson equation of Computer Exercise 4.1 by extending the function `Compute_Mixed` of Section 4.8.3. Verify your program by solving a test problem.

COMMENTS AND REFERENCES

A comprehensive discussion of second-order elliptic, parabolic, and hyperbolic PDEs, their weak formulations, and the existence and the regularity of their solutions can be found in [7, Chapters 6 and 7]. Weak formulations of elliptic and time-dependent PDEs subject to various boundary conditions are also studied in [6, Chapters 3 and 6].

The proof of the Lax-Milgram lemma can be found in [5, Section 1.1]. This result is only a sufficient condition for the well-posedness of the abstract problem (4.18). A more general condition which is both necessary and sufficient for the well-posedness of (4.18) is given in [6, Section 2.1]. This condition is expressed in terms of the so-called inf-sup conditions. The ellipticity of the weak formulations of Section 4.2 for the PDE (4.1) subject to different boundary conditions is proved in [6, Section 3.1.2]. Other examples of elliptic boundary value problems including linearized elasticity and the fourth-order biharmonic problem are studied in [5, Section 1.2].

It is well-known that discretizations of well-posed problems may not be well-posed. Discretizations of mixed formulations are classical examples in this regard. In [6, Section 2.4], necessary and sufficient conditions for the well-posedness of a class of saddle-point problems and their Galerkin discretizations are presented.

As discussed in [5, Section 2.4], Céa's lemma is the simplest convergence result for the abstract problem (4.18). In practice, rather than (4.23), Galerkin approximations of (4.18) lead to discrete problems of the type

$$A_h(u_h, v_h) = F_h(v_h), \text{ for all } v_h \in X_h,$$

where A_h and F_h are mesh-dependent approximations of A and F. A typical example is the effect of numerical integration where one employs numerical integration methods such as quadrature schemes to evaluate $A(u, v)$ and $F(v)$. Other common examples are non-conforming methods and domains with curved boundaries. These cases are special instances of *variational crimes*. A convergence analysis suitable for variational crimes including the Strang lemmas are discussed in [5, Chapter 4]. A more general convergence result is given in [6, Section 2.3]. Roughly speaking, this result states that stability, consistency, and approximability lead to convergence, where stability means well-posedness of discrete problems with mesh-independent well-posedness parameters and consistency means the exact solution satisfies the discrete problem as $h \to 0$.

A comprehensive study of finite element approximations of specific classes of time-dependent PDEs by using the method of lines is provided in [6, Chapter 6]. In particular, stability and convergence of finite element methods for parabolic problems using implicit Euler, explicit Euler, and the backward difference formula of second order is studied in [6, Section 6.1.6].

The term mixed formulation usually refers to problems associated to a saddle-point of a functional such as the Stokes problem [4, Chapter 5]. Sometimes it is also used for formulations in which unknowns involve a function together with some of its derivatives [5, Page 417]. A comprehensive reference for mixed finite element methods and inf-sup conditions is [4]. A more concise reference for mixed methods is [6, Chapter 4]. To computationally verify inf-sup conditions, one can employ the framework discussed in [4, Section 3.4.3] for verification of inf-sup conditions using singular-values. Computational verification of inf-sup conditions using FEniCS is discussed in [10, Chapter 36].

Several examples for solving PDEs using FEniCS and the availble options for FEniCS functions and objects can be found in the FEniCS Project webpage `https://fenicsproject.org/`, the FEniCS Tutorial [9], and the FEniCS Book [10]. Although some parts are outdated, the first Chapter of the FEniCS Book [10] provides an excellent overview about using FEniCS to solve different types of boundary value problems. More detailed discussions on the capabilities of UFL for defining different types of problems can be found in the FEniCS Project webpage and the FEniCS Book [10, Chapter 17].

5 Applications

A selection of applications in civil and mechanical engineering is studied in this chapter and Python programs for their implementation are provided. These applications are selected from structural engineering, elasticity, vibrations, fluid mechanics, heat transfer, pavement engineering, and geotechnical engineering. Some of these applications are a special case of the general classes of PDEs discussed in Chapter 4 while others including elasticity and the Stokes equation may be considered as generalizations of those classes. The Euler-Bernoulli beam theory is the only application involving a fourth-order differential equation. Although nonlinear problems are beyond the purview of this book, we also consider nonlinear elasticity to provide an application for the Nédélec finite element.

As these applications involve standard topics, we mainly focus on their implementation and refer the readers to the standard texts in the associated field for more discussions on the derivation of governing equations based on the underlying physical principles. Python codes for all applications are provided in each section as well as the companion website of the book. Unlike Chapter 4, we do not provide a detailed dissection of these codes as we assume the readers have carefully examined the codes of Chapter 4 and are already able to understand the main structure of these codes.

5.1 ELASTIC BARS

We compute deformations of an elastic bar under axial loads. Let $\Omega = (0, l)$ represent a linearly elastic bar of length l with the cross section area $A(x)$ at $x \in [0, l]$ and Young's modulus E. We assume that the bar is fixed at $x = 0$, that an axial load f is exerted at $x = l$, and that a distributed axial body load with the density $b(x)$ per unit length is applied to the bar, see Figure 5.1.

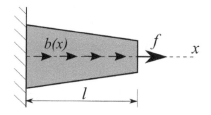

Figure 5.1 An elastic bar under a distributed load $b(x)$ and an axial point load f.

The equilibrium of the bar in the x-direction implies that the displacement $u(x)$ at $x \in \Omega$ in the x-direction is the solution of the following boundary value problem:

Find $u(x)$ such that

$$\begin{cases} -\partial_x(EA\partial_x u) = b, \text{ in } \Omega, \\ u(0) = 0, \text{ and } \partial_x u|_{x=l} = \frac{f}{EA(l)}. \end{cases} \qquad (5.1)$$

This problem is an elliptic equation with a mixed Dirichlet-Neumann boundary condition discussed in Section 4.2. Therefore, a weak formulation for this problem can be stated as: Find $u \in H_D^1(\Omega)$ such that

$$\int_0^l EA\partial_x u \, \partial_x v = \int_0^l bv + fv(l), \text{ for all } v \in H_D^1(\Omega), \qquad (5.2)$$

where $H_D^1(\Omega) = \{v \in H^1(\Omega) : v(0) = 0\}$. Let $X_h \subset H_D^1(\Omega)$ be any H^1-conformal finite element space associated to a mesh of Ω. Then, a conformal finite element method for (5.2) reads: Find $u_h \in X_h$ such that

$$\int_0^l EA\partial_x u_h \partial_x v_h = \int_0^l bv_h + fv_h(l), \text{ for all } v_h \in X_h. \qquad (5.3)$$

Example 5.1. If a solution u of (5.2) belongs to X_h, then u is also a solution of (5.3). The uniqueness of the solution of (5.3) then implies that $u_h = u$ regardless of the refinement level of the underlying mesh of X_h. As an example, let $b = 0$, and assume that the bar has a constant cross section area A. Then, the solution of 5.1 simply reads $u(x) = \frac{f}{EA}x$. Now let T_h be a mesh of Ω consisting of only one element $K = [0, l]$ and suppose X_h is the finite element space associated to 1-simplex of type (1). Then, $u \in X_h$. It is easy to show that $\dim X_h = 1$, and the only global shape function is $\psi_1(x) = \frac{x}{l}$. Let $u_h(x) = C\psi_1(x)$. By using $v_h = \psi_1$, the finite element method (5.3) yields

$$\int_0^l EA\frac{C}{l} \cdot \frac{1}{l} dx = f,$$

which means $C = \frac{fl}{EA}$. Consequently,

$$u_h(x) = C\psi_1(x) = \frac{fx}{EA} = u(x).$$

Therefore, the approximate solution u_h is equal to the exact solution although the underlying mesh consists of only one element.

Example 5.2. In (5.1), suppose EA is constant and consider the non-dimensionalized variables

$$\tilde{x} = \frac{x}{l}, \ \tilde{u} = \frac{u}{l}, \ \tilde{b} = \frac{l}{EA}b, \ \tilde{f} = \frac{1}{EA}f.$$

Let $\tilde{\Omega}$ be the unit interval. Then, the weak problem (5.2) in terms of the above variables reads: Find $\tilde{u} \in H_D^1(\tilde{\Omega})$ such that

$$\int_0^1 \partial_{\tilde{x}}\tilde{u} \, \partial_{\tilde{x}}\tilde{v} = \int_0^1 \tilde{b}\tilde{v} + \tilde{f}\tilde{v}(1), \text{ for all } \tilde{v} \in H_D^1(\tilde{\Omega}),$$

where $H_D^1(\tilde{\Omega}) = \{\tilde{v} \in H^1(\tilde{\Omega}) : \tilde{v}(0) = 0\}$. Assuming $\tilde{f} = 0$, and $\tilde{b}(\tilde{x}) = \frac{\pi^2}{4} \sin \frac{\pi}{2} \tilde{x}$, the solution of the non-dimensionalized problem simply reads $\tilde{u}_e(\tilde{x}) = \sin \frac{\pi}{2} \tilde{x}$. The following program employs 1-simplex of type (k) to compute the L^2-error associated to a uniform mesh of $\tilde{\Omega}$ with N elements.

```
def Compute_Bar(N, k):
    """ Given mesh division size N & the degree k of Lagrange
    elements, this function returns the mesh size h and error
    of the finite element solution of an elastic bar """

    # create mesh and define function space
    mesh = UnitIntervalMesh(N)
    V_h = FunctionSpace(mesh, "Lagrange", k)

    # the exact solution
    u_e = Expression('sin(0.5*pi*x[0])', degree = 3 + k)

    # define Dirichlet boundary condition
    def left_boundary(x, on_boundary):
        tol = 1E-12    # tolerance for comparisons
        return on_boundary and abs(x[0]) < tol
    BC = DirichletBC(V_h, Constant(0.0), left_boundary)

    # define the weak formulation
    u_h = TrialFunction(V_h); v_h = TestFunction(V_h)
    b = Expression('pi*pi*sin(0.5*pi*x[0])/4.0', degree= 3+k)
    A = inner(grad(u_h), grad(v_h))*dx
    F = b*v_h*dx

    # compute finite element solution
    u_h = Function(V_h)         # u_h is redefined
    solve(A == F, u_h, BC)
    L2_Error = errornorm(u_e, u_h, norm_type="L2")

    return L2_Error, mesh.hmax()
```

By using this function, we can compute convergence rates. For example, employing 1-simplex of type (1) with a family of meshes with $N = 3, 6, 9$, yields:

```
N =  3, L2_error = 1.76E-02
N =  6, L2_error = 4.42E-03
N =  9, L2_error = 1.96E-03
L2-convergence rate = 2.00
```

This convergence rate is the optimal convergence rate for 1-simplex of type (1).

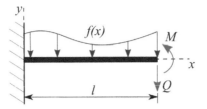

Figure 5.2   A beam under a distributed load $f(x)$, a shear force Q, and a bending moment M.

5.2   EULER-BERNOULLI BEAMS

Let $\Omega = (0, l)$ represent a cantilever beam of length l shown in Figure 5.2 which is fixed at $x = 0$ with the flexural rigidity $EI(x)$, $x \in \Omega$. We assume that the beam is under a distributed load $f(x)$ per unit length, and a shear force Q and a bending moment M at $x = l$. *The Euler-Bernoulli beam theory* states that the deflection $u(x)$ of the beam in the y-direction satisfies the following fourth-order boundary value problem:

$$\begin{cases} \partial_{xx}(EI\partial_{xx}u) = f, & \text{in } \Omega, \\ u(0) = 0, & \partial_x u|_{x=0} = 0, \\ (EI\partial_{xx}u)|_{x=l} = M, & \partial_x(EI\partial_{xx}u)|_{x=l} = Q. \end{cases} \tag{5.4}$$

To obtain a weak formulation for (5.4), we multiply both sides of the governing equation by an arbitrary test function v and integrate to obtain

$$\int_0^l v\partial_{xx}(EI\partial_{xx}u) = \int_0^l fv. \tag{5.5}$$

As we want to use the same solution and test spaces, we should derive a weak form with the same highest order of derivatives of u and v. This can be achieved by twice applying Green's formula (also called the integration by parts in 1D) to the left side of (5.5). More specifically, we can write

$$\int_0^l v\partial_{xx}(EI\partial_{xx}u) = -\int_0^l \partial_x(EI\partial_{xx}u)\partial_x v + \left(v\partial_x(EI\partial_{xx}u)\right)\big|_0^l$$

$$= \int_0^l EI\partial_{xx}u\,\partial_{xx}v - (EI\partial_{xx}u\,\partial_x v)\big|_0^l + \left(v\partial_x(EI\partial_{xx}u)\right)\big|_0^l.$$

Recall that $H^2(\Omega)$ is the space of L^2 functions on Ω with L^2 first- and second-order derivatives, see Example 2.8. By assuming $v(0) = \partial_x v(0) = 0$, and using the above relation and the boundary conditions, one obtains the following weak formulation for (5.4): Find $u \in H_D^2(\Omega)$ such that

$$\int_0^l EI\partial_{xx}u\,\partial_{xx}v = \int_0^l fv + M\partial_x v(l) - Qv(l), \text{ for all } v \in H_D^2(\Omega), \tag{5.6}$$

where $H_D^2(\Omega) = \{v \in H^2(\Omega) : v(0) = \partial_x v(0) = 0\}$.

Suppose $X_h \subset H_D^2(\Omega)$ is an H^2-conformal finite element space associated to a family of Hermite 1-simplex of type (3) introduced in Section 3.2.2. Then, a conforming finite element method for (5.6) can be stated as: Find $u_h \in X_h$ such that

$$\int_0^l EI \partial_{xx} u_h \, \partial_{xx} v_h = \int_0^l f v_h + M \partial_x v_h(l) - Q v_h(l), \text{ for all } v_h \in X_h. \qquad (5.7)$$

Example 5.3. Suppose that EI is constant, $f = 0$, and $M = 0$. Then, it is easy to show that the solution of (5.4) is

$$u(x) = \frac{Q}{6EI} x^2 (x - 3l). \qquad (5.8)$$

Since this solution is a polynomial of degree 3 belonging to X_h, the solution u_h of (5.7) with the data of this example will be equal to the exact solution regardless of the underlying mesh. For example, let T_h be a mesh consisting of only 1 element $K = [0, l]$. The local shape functions (3.12) imply that the global shape functions of the associated finite element space with no Dirichlet boundary conditions are

$$\theta_0(x) = \frac{1}{l^3}(x - l)^2(2x + l), \quad \theta_1(x) = \frac{1}{l^2}x(x - l)^2,$$

$$\theta_2(x) = \frac{1}{l^3}x^2(3l - 2x), \quad \theta_3(x) = \frac{1}{l^2}x^2(x - l).$$

The finite element method (5.7) then simply reads: Find $u_h \in X_h$ such that

$$\int_0^l \partial_{xx} u_h \, \partial_{xx} v_h = -\frac{Q}{EI} v_h(l), \text{ for all } v_h \in X_h, \qquad (5.9)$$

where $\dim X_h = 2$, and $\{\theta_2, \theta_3\}$ are the global shape functions of X_h. Since $u_h \in X_h$, we can write $u_h = U_2 \theta_2 + U_3 \theta_3$, where U_2 and U_3 are unknown constants. Substituting this relation into (5.9) and assuming $v_h = \theta_2$, and $v_h = \theta_3$, yield

$$\int_0^l (U_2 \partial_{xx} \theta_2 + U_3 \partial_{xx} \theta_3) \partial_{xx} \theta_2 = -\frac{Q}{EI},$$

$$\int_0^l (U_2 \partial_{xx} \theta_2 + U_3 \partial_{xx} \theta_3) \partial_{xx} \theta_3 = 0.$$

These equations can be stated in the matrix form $\mathbb{A} \cdot \mathbb{U} = \mathbb{F}$, where $\mathbb{U}^T = (U_2, U_3)$, $\mathbb{F}^T = (-\frac{Q}{EI}, 0)$, and the stiffness matrix $\mathbb{A}$ is given by

$$\mathbb{A} = \begin{bmatrix} \int_0^l \partial_{xx} \theta_2 \partial_{xx} \theta_2 & \int_0^l \partial_{xx} \theta_2 \partial_{xx} \theta_3 \\ \int_0^l \partial_{xx} \theta_2 \partial_{xx} \theta_3 & \int_0^l \partial_{xx} \theta_3 \partial_{xx} \theta_3 \end{bmatrix} = \begin{bmatrix} 12/l^3 & -6/l^2 \\ -6/l^2 & 4/l \end{bmatrix}.$$

Solving the above linear system results in $U_2 = -\frac{Ql^3}{3EI}$, and $U_3 = -\frac{Ql^2}{2EI}$. Therefore,

$$u_h(x) = U_2 \theta_2(x) + U_3 \theta_3(x) = \frac{Q}{6EI} x^2 (x - 3l),$$

which is equal to the exact solution (5.8) as we expected.

Currently, the Hermite element is not fully supported in FEniCS and hence, it is not possible to directly implement the conformal finite element method (5.7) in FEniCS. To overcome this issue, we can employ a mixed formulation for the beam problem by considering the bending moment $m = EI\partial_{xx}u$ as a separate unknown. More specifically, our goal is to find the deflection $u(x)$ and the bending moment $m(x)$ such that

$$
\begin{cases}
\partial_{xx}m = f, & \text{in } \Omega, \\
\partial_{xx}u - \frac{1}{EI}m = 0, & \text{in } \Omega, \\
u(0) = 0, & \partial_x u(0) = 0, \\
m(l) = M, & \partial_x m(l) = Q.
\end{cases} \tag{5.10}
$$

Let $H_L^1(\Omega) = \{v \in H^1(\Omega) : v(0) = 0\}$, and $H_R^1(\Omega) = \{v \in H^1(\Omega) : v(l) = 0\}$. By using the integration by parts, it is straightforward to write the following mixed formulation based on (5.10): Find $u, m \in H^1(\Omega)$ with $u(0) = 0$, and $m(l) = M$ such that

$$
\begin{aligned}
\int_0^l \partial_x m\, \partial_x v &= -\int_0^l fv + Qv(l), & \text{for all } v \in H_L^1(\Omega), \\
\int_0^l \partial_x u\, \partial_x r &+ \int_0^l \frac{1}{EI}mr = 0, & \text{for all } r \in H_R^1(\Omega).
\end{aligned} \tag{5.11}
$$

Let $\bar{L}_h$ and $\bar{R}_h$ be the finite element spaces associated to families of 1-simplices of type (k) and $(\hat{k})$, respectively. Also suppose $L_h = \{v \in \bar{L}_h : v(0) = 0\}$, and $R_h = \{v \in \bar{R}_h : v(l) = 0\}$. Then, a conformal mixed finite element method based on (5.11) can be stated as: Find $u_h \in \bar{L}_h$ and $m_h \in \bar{R}_h$ with $u_h(0) = 0$, and $m_h(l) = M$ such that

$$
\begin{aligned}
\int_0^l \partial_x m_h\, \partial_x v_h &= -\int_0^l fv_h + Qv_h(l), & \text{for all } v_h \in L_h, \\
\int_0^l \partial_x u_h\, \partial_x r_h &+ \int_0^l \frac{1}{EI}m_h r_h = 0, & \text{for all } r_h \in R_h.
\end{aligned} \tag{5.12}
$$

Let $\dim L_h = D$ and $\dim R_h = B$ with the global shape functions $\{\phi_i\}_{i=1}^D$ and $\{\psi_i\}_{i=1}^B$, respectively. It is straightforward to show that the stiffness matrix of the mixed finite element method (5.12) is of the form

$$
\begin{bmatrix}
\mathbf{0} & \mathbf{S}_{D \times B} \\
\hline
\mathbf{S}_{B \times D}^T & \bar{\mathbf{S}}_{B \times B}
\end{bmatrix},
$$

with $\mathbf{S}_{ij} = \int_0^l \partial_x \phi_i\, \partial_x \psi_j$, and $\bar{\mathbf{S}}_{ij} = \int_0^l \psi_i\, \psi_j$. Following the discussion of Section 4.8.2, we conclude that if this stiffness matrix is invertible then the submatrix $\mathbf{S}$ is full rank. Therefore, the above mixed finite element method is not stable if $B < D$, or equivalently, if the degree of the finite element for m_h is smaller than that of u_h. For example, the choice of simplicial elements with $k = 2$ and $\hat{k} = 1$ is not stable.

Example 5.3 can be solved by using the following function:

```
def Compute_MixedBeam(N, k_u, k_m):
    """ Given mesh division size N, and the degrees k_u and
    k_m of Lagrange elements for deflection and moment, this
    function returns the mesh size h and errors of a mixed
    FEM for beams """

    # create mesh and define function space
    mesh = IntervalMesh(N, 0.0, 1)
    LGu = FiniteElement("Lagrange", mesh.ufl_cell(), k_u)
    LGm = FiniteElement("Lagrange", mesh.ufl_cell(), k_m)
    Z_h = FunctionSpace(mesh, MixedElement([LGu, LGm]))

    # the exact solution
    u_e = Expression('Q*x[0]*x[0]*(x[0] - 3*l)/(6*EI)',
            l = 1, EI = EI, Q = Q, degree = 3 + k_u)
    m_e = Expression('Q*(x[0] - 1)', l = 1, Q = Q,
                    degree = 3 + k_m)

    # marking the boundary using a mesh function
    boundary_parts = MeshFunction("size_t", mesh,
                                    mesh.topology().dim()-1)

    # mark right boundary edges as subdomain 1
    class Right_boundary(SubDomain):
        def inside(self, x, on_boundary):
            tol = 1E-12   # tolerance for comparisons
            return on_boundary and abs(x[0] - 1) < tol
    Gamma_R = Right_boundary()
    Gamma_R.mark(boundary_parts, 1)

    # define Dirichlet boundary condition
    def Left(x, on_boundary):
        tol = 1E-12   # tolerance for comparisons
        return on_boundary and abs(x[0]) < tol

    def Right(x, on_boundary):
        tol = 1E-12   # tolerance for comparisons
        return on_boundary and abs(x[0] - 1) < tol

    bcs = [DirichletBC(Z_h.sub(0), Constant(0.0), Left),
            DirichletBC(Z_h.sub(1), Constant(M), Right)]

    # define the weak formulation
    (u_h, m_h) = TrialFunctions(Z_h)    # trial functions
```

```
(v_h, r_h) = TestFunctions(Z_h)      # test fucntions
LH = ( m_h.dx(0)*v_h.dx(0) + u_h.dx(0)*r_h.dx(0)
        + Constant(1.0/EI)*m_h*r_h )*dx    # left side
f = Constant(0.0)                    # no distributed load
ds = Measure("ds", domain=mesh,
   subdomain_data=boundary_parts)    # boundary integral
RH = -f*v_h*dx + Constant(Q)*v_h*ds(1)          # right side


# compute finite element solution
z_h = Function(Z_h)
solve(LH == RH, z_h, bcs)
(u_h, m_h) = z_h.split()


Error_u = errornorm(u_e, u_h, norm_type="L2")
Error_m = errornorm(m_e, m_h, norm_type="L2")


return Error_u, Error_m, mesh.hmax()
```

Notice that in the definition of the weak formulation, m_h.dx(0) represents $\partial_x m_h$. In general, f.dx(i) denotes the partial derivative of f with respect to x[i]. As an example, using elements of degree 1 for both u_h and m_h, the mesh sizes $N = 3, 6, 9$, and the data $EI = 1, l = 1, f = 0, M = 0$, and $Q = 1$ result in the following L^2-errors and the convergence rate of u_h:

```
N =  3, error_u = 5.78E-03, error_m = 3.03E-16
N =  6, error_u = 1.46E-03, error_m = 1.73E-16
N =  9, error_u = 6.50E-04, error_m = 1.50E-16
rate_u = 2.00
```


5.3   ELASTIC MEMBRANES

The boundary value problem for the equilibrium position of an elastic membrane fixed at its boundary and under a vertical load f is given in (1.1). This problem is simply Poisson's equation with the homogeneous Dirichlet boundary condition. The associated weak formulation is (4.7), where $b = 0$, $c = 0$, and $\mathbb{D}$ is the identity matrix $\mathbb{I}$.

The implementation of this problem in FEniCS is straightforward. For example, the following program computes the vertical displacement of the unit square membrane under a constant load shown in Figure 1.4.

```
# create mesh and define function space
mesh = UnitSquareMesh(8, 8)
V = FunctionSpace(mesh, 'CG', 1)


# define boundary condition
def boundary(x, on_boundary):
    return on_boundary
```

```
bc = DirichletBC(V, Constant(0.0), boundary)

# define variational problem
u = TrialFunction(V)
w = TestFunction(V)
f = Constant(4.0)
a = dot(grad(u), grad(w))*dx
L = f*w*dx

# compute solution
u = Function(V)
solve(a == L, u, bc)
```

5.4   THE WAVE EQUATION

The wave equation is used to model a wide range of phenomena including vibrations of strings and membranes, underwater sound wave propagation, and magnetic waves in the sun's surface. Here, we solve the wave equation to model a traveling wave in a rectangular membrane.

More specifically, let $\Omega = (0,L) \times (0,W)$ denote a rectangular membrane with the length L and the width W and suppose $u(x,y,t)$ is the displacement of the point (x,y) at time t in the z-direction. Let $\Gamma_D = \Gamma_L \cup \Gamma_R$ be the union of the left boundary Γ_L and the right boundary Γ_R of Ω and let Γ_N denote the top and the bottom boundary of Ω. We fix the membrane at Γ_R and create a traveling wave in Ω propagating in the x-direction by imposing a time-dependent boundary condition at Γ_L. We also assume the normal derivative of u at Γ_N vanishes. To model the wave propagation in the membrane, we solve the following initial-boundary value problem: Find $u(x,y,t)$ such that

$$\begin{cases} \partial_{tt}u - C^2\Delta u = 0, & \text{in } \Omega, \ t \in [0,T], \\ u = g, & \text{on } \Gamma_L, \ t \in [0,T], \\ u = 0, & \text{on } \Gamma_R, \ t \in [0,T], \\ \partial_y u = 0, & \text{on } \Gamma_N, \ t \in [0,T], \\ u = \partial_t u = 0, & \text{at } t = 0, \end{cases} \tag{5.13}$$

where C is a constant depending on physical properties of Ω. The function g is defined as

$$g(t) = \begin{cases} A\sin^2 \omega t, & \text{if } 0 \leq t \leq \frac{\pi}{\omega}, \\ 0, & \text{if } t > \frac{\pi}{\omega}, \end{cases}$$

where A and ω are constants.

A weak formulation for (5.13) can be stated as: Find $u(x,y,t)$ satisfying $u(\cdot,t) \in H^1(\Omega)$ for any $t \in [0,T]$, with $u = g$, on Γ_L, and $u = 0$, on Γ_R, such that

$$\begin{cases} \int_\Omega v\partial_{tt}u + \int_\Omega C^2\nabla u \cdot \nabla v = 0, & \text{for all } v \in H_D^1(\Omega), \\ u = \partial_t u = 0, & \text{at } t = 0. \end{cases} \tag{5.14}$$

Let $V_h \subset H^1(\Omega)$ be a finite element space and let $X_h = V_h \cap H_D^1(\Omega)$. By using the three-point centered-difference formula mentioned in Exercise 4.19 with the uniform time increment $S = t_{i+1} - t_i$, we can write the following implicit method for (5.14): Given $u_h^i, u_h^{i+1} \in V_h$, find $u_h^{i+2} \in V_h$ with $u_h^{i+2} = g(t_{i+2})$, on Γ_L, and $u_h^{i+2} = 0$, on Γ_R, such that

$$\int_\Omega u_h^{i+2} v_h + S^2 C^2 \int_\Omega \nabla u_h^{i+2} \cdot \nabla v_h = 2 \int_\Omega u_h^{i+1} v_h - \int_\Omega u_h^i v_h, \text{ for all } v_h \in X_h, \quad (5.15)$$

with $u_h^0 = u_h^1 = 0$. Notice that the wave equation is a hyperbolic PDE defined in Exercise 4.17. The finite element method (5.15) can be implemented in FEniCS as follows.

```
# parameters
C = 1.0                    # membrane's constant
T = 10                     # final time
num_steps = 400            # number of time steps
S = T / num_steps          # time step size
L = 4                      # length
W = 0.5                    # width
t = 0.0                    # initial time
A = 10.0                   # the amplitude
omega = 0.3                # the frequency
nx = 400 ;ny = 50          # number of divisions
mesh = RectangleMesh(Point(0.0, 0.0), Point(L, W), nx, ny)
V = FunctionSpace(mesh, 'CG', 1)

value_0 = Constant((0.0))
# defining u at t=0 as u0 and t=S as u1
u0 = interpolate(value_0, V)
u1 = interpolate(value_0, V)

# defining variational problem
u = TrialFunction(V)
v = TestFunction(V)
a = u*v*dx + S*S*C*C*inner(grad(u), grad(v))*dx
l = 2*u1*v*dx-u0*v*dx

# time-dependent boundary condition
g = Expression('A*pow(sin(omega*t),2)', A = A,
                omega = omega, t = 0.0, degree = 4)

# defining boundaries
def left(x, on_boundary):
    tol = 1E-12    # tolerance for coordinate comparisons
    return on_boundary and abs(x[0]) < tol
```

```
def right(x, on_boundary):
    tol = 1E-12    # tolerance for coordinate comparisons
    return on_boundary and abs(x[0] - L) < tol

# create VTK file for saving solution
vtkfile = File("Ch5_Wave/Wave_disp.pvd")

u=Function(V)
# time-stepping to solve time-dependent BVP
while t <= T:
    g.t = t     # updating time parameter

    if t <= pi/omega:
        bc_l = DirichletBC(V, g, left)
    else:
        bc_l = DirichletBC(V, Constant(0.0), left)
    bc_r = DirichletBC(V, Constant(0.0), right)
    bc = [bc_l, bc_r]

    A, b = assemble_system(a, l, bc)
    solve(A, u.vector(), b)

    # save to file
    vtkfile << (u,t)

    # updating previous solution
    u0.assign(u1)
    u1.assign(u)

    # updating time
    t += dt
```

By using ParaView, it is straightforward to create an animation from the output of this code. Figure 5.3 depicts configurations of the membrane at different time steps corresponding to the values of parameters given in the above code.

5.5   HEAT TRANSFER IN A TURBINE BLADE

We compute the transient temperature distribution $u(x,y,t)$ in a turbine blade Ω of a jet engine depicted in Figure 5.4. The bottom boundary Γ_N^1 is exposed to hot gases from combustor while the top boundary Γ_N^2 is exposed to cooling air. We solve the

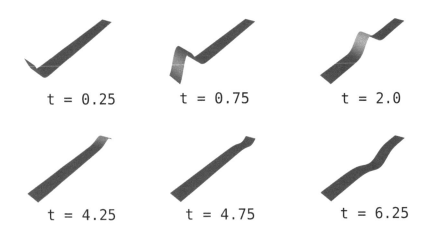

Figure 5.3 Configurations of a vibrating membrane at different time steps.

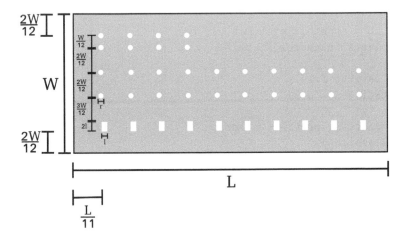

Figure 5.4 Geometry of a turbine blade of a jet engine with cooling film holes.

following advection-diffusion equation to determine u:

$$\begin{cases} \partial_t u - \mathrm{div}(\varepsilon \nabla u) + \boldsymbol{\beta} \cdot \nabla u = 0, & \text{in } \Omega,\ t \in [0,T], \\ \varepsilon \partial_y u = Q_1(t), & \text{on } \Gamma_N^1,\ t \in [0,T], \\ \varepsilon \partial_y u = Q_2(t), & \text{on } \Gamma_N^2,\ t \in [0,T], \\ \boldsymbol{n} \cdot \nabla u = 0, & \text{on } \partial\Omega - (\Gamma_N^1 \cup \Gamma_N^2),\ t \in [0,T], \\ u = u_0, & \text{at } t = 0, \end{cases}$$

where $\boldsymbol{n}$ is the unit normal vector field of $\partial\Omega$, $\varepsilon(x,y)$ is the diffusion coefficient, $\boldsymbol{\beta}$ is a given velocity field, and $Q_1(t)$ and $Q_2(t)$ are known scalar-valued functions. A weak form for this initial-boundary value problem can be stated as: Find $u(x,y,t)$ with $u(\cdot,t) \in H^1(\Omega)$ for any $t \in [0,T]$, such that

$$
\begin{cases}
\int_\Omega v \partial_t u + \int_\Omega \varepsilon \nabla u \cdot \nabla v + \int_\Omega (\boldsymbol{\beta} \cdot \nabla u) v = \sum_{j=1}^2 \int_{\Gamma_N^j} Q_j v, & \text{for all } v \in H^1(\Omega), \\
u = u_0, & \text{at } t = 0.
\end{cases}
$$

Suppose $V_h \subset H^1(\Omega)$ is a finite element space. Discretizing the above weak formulation by using V_h and the backward Euler method with the uniform time increment $S = t_{i+1} - t_i$, yields the following implicit finite element method: Given $u_h^i \in V_h$, find $u_h^{i+1} \in V_h$ such that

$$
\int_\Omega u_h^{i+1} v_h + S \int_\Omega \varepsilon \nabla u_h^{i+1} \cdot \nabla v_h + S \int_\Omega (\boldsymbol{\beta} \cdot \nabla u_h^{i+1}) v_h =
$$

$$
S \sum_{j=1}^2 \int_{\Gamma_N^j} Q_j v_h + \int_\Omega u_h^i v_h, \text{ for all } v_h \in V_h,
$$

where u_h^0 is the interpolation of u_0 in V_h. This finite element method can be implemented using FEniCS as follows.

```
# parameters
T = 0.35                # final time
t = 0.0                 # initial time
num_steps = 1000        # number of time steps
S = T / num_steps       # time step size
L = 0.6 ; W = 0.3       # length and width of the blade
N = 20                  # mesh resolution
degreeCG = 2            # FE degree
eps = 0.1               # diffusion coefficient
r = 0.005               # circles radius
l = 0.01                # small rec length


# generating mesh
blade = Rectangle(Point(0.0, 0.0), Point(L, W))
cyc11 = Circle(Point(L/11,5*W/12), r)
cyc12 = Circle(Point(2*L/11,5*W/12), r)
cyc13 = Circle(Point(3*L/11,5*W/12), r)
cyc14 = Circle(Point(4*L/11,5*W/12), r)
cyc15 = Circle(Point(5*L/11,5*W/12), r)
cyc16 = Circle(Point(6*L/11,5*W/12), r)
cyc17 = Circle(Point(7*L/11,5*W/12), r)
cyc18 = Circle(Point(8*L/11,5*W/12), r)
cyc19 = Circle(Point(9*L/11,5*W/12), r)
cyc110 = Circle(Point(10*L/11,5*W/12), r)
cyc21 = Circle(Point(L/11,7*W/12), r)
cyc22 = Circle(Point(2*L/11,7*W/12), r)
cyc23 = Circle(Point(3*L/11,7*W/12), r)
```

```
cyc24 = Circle(Point(4*L/11,7*W/12), r)
cyc25 = Circle(Point(5*L/11,7*W/12), r)
cyc26 = Circle(Point(6*L/11,7*W/12), r)
cyc27 = Circle(Point(7*L/11,7*W/12), r)
cyc28 = Circle(Point(8*L/11,7*W/12), r)
cyc29 = Circle(Point(9*L/11,7*W/12), r)
cyc210 = Circle(Point(10*L/11,7*W/12), r)
cyc31 = Circle(Point(L/11,9*W/12), r)
cyc32 = Circle(Point(2*L/11,9*W/12), r)
cyc33 = Circle(Point(3*L/11,9*W/12), r)
cyc34 = Circle(Point(4*L/11,9*W/12), r)
cyc41 = Circle(Point(L/11,10*W/12), r)
cyc42 = Circle(Point(2*L/11,10*W/12), r)
cyc43 = Circle(Point(3*L/11,10*W/12), r)
cyc44 = Circle(Point(4*L/11,10*W/12), r)
rec1 = Rectangle(Point(L/11, 2*W/12), \
                 Point(1+ L/11, 2*1+W/6))
rec2 = Rectangle(Point(2*L/11, 2*W/12), \
                 Point(1+ (2*L/11), 2*1+W/6))
rec3 = Rectangle(Point(3*L/11, 2*W/12), \
                 Point(1+ (3*L/11), 2*1+W/6))
rec4 = Rectangle(Point(4*L/11, 2*W/12), \
                 Point(1+ (4*L/11), 2*1+W/6))
rec5 = Rectangle(Point(5*L/11, 2*W/12), \
                 Point(1+ (5*L/11), 2*1+W/6))
rec6 = Rectangle(Point(6*L/11, 2*W/12), \
                 Point(1+ (6*L/11), 2*1+W/6))
rec7 = Rectangle(Point(7*L/11, 2*W/12), \
                 Point(1+ (7*L/11), 2*1+W/6))
rec8 = Rectangle(Point(8*L/11, 2*W/12), \
                 Point(1+ (8*L/11), 2*1+W/6))
rec9 = Rectangle(Point(9*L/11, 2*W/12), \
                 Point(1+ (9*L/11), 2*1+W/6))
rec10 = Rectangle(Point(10*L/11, 2*W/12), \
                 Point(1+ (10*L/11), 2*1+W/6))
domain = blade - cyc11 - cyc12 - cyc13 - cyc14\
 - cyc15 - cyc16 - cyc17 - cyc18 - cyc19 - cyc110\
 - cyc21 - cyc22 - cyc23 - cyc24 - cyc25 - cyc26\
 - cyc27 - cyc28 - cyc29 - cyc210 - cyc31 - cyc32\
 - cyc33 - cyc34 - cyc41 - cyc42 - cyc43 - cyc44\
 - rec1 - rec2 - rec3 - rec4 - rec5 - rec6 - rec7\
 - rec8 - rec9 - rec10
mesh = generate_mesh(domain, N)
```

```
# symbolic physical coordinates for given mesh
x = SpatialCoordinate(mesh)

# define function space
V = FunctionSpace(mesh, 'CG', degreeCG)

# define boundaries
boundary_parts = MeshFunction('size_t', mesh, \
                              mesh.topology().dim()-1)

class BottomBoundary(SubDomain):
    def inside(self, x, on_boundary):
        tol = 1E-14   # tolerance for coordinate comparisons
        return on_boundary and abs(x[1]) < tol

Gamma_1 = BottomBoundary()
Gamma_1.mark(boundary_parts, 1)

class TopBoundary(SubDomain):
    def inside(self, x, on_boundary):
        tol = 1E-14   # tolerance for coordinate comparisons
        return on_boundary and abs(x[1] - W) < tol

Gamma_1 = TopBoundary()
Gamma_1.mark(boundary_parts, 2)

# define initial value
T0 = Constant(0.0)
u_n = interpolate(T0, V)

# define variational problem
ds = Measure("ds", domain=mesh, \
             subdomain_data=boundary_parts)
u = TrialFunction(V)
v = TestFunction(V)
beta = as_vector([4*x[1]*(1-x[1]),0.0])
Q1 = Expression(('400*t'), t=t, degree=1)
Q2 = Expression(('-20*t'), t=t, degree=1)
a = (u*v +S*eps*inner(grad(u),grad(v))
+ S*inner(beta,grad(u))*v)*dx
L = S*eps*dot(Q1,v)*ds(1)+S*eps*dot(Q2,v)*ds(2)+u_n*v*dx

# produce a vtkfile for visualization
vtkfile = File('Ch5_heat/ConvDiff.pvd')
```

```
# solver
u = Function(V)
for n in range(num_steps):

    # update current time
    t += S

    # compute solution
    solve(a == L, u)

    # save solution to vtk format
    vtkfile << (u,t)

    # update tractions
    Q1.t = t
    Q2.t = t

    # update previous solution
    u_n.assign(u)
```

Figure 5.5 shows the distribution of temperature at some time steps. These results are computed by using the parameters and input functions given in the above program.

5.6   SEEPAGE IN EMBANKMENT

Seepage problems deal with flow of fluids through porous media. A standard example is seepage in dams or sheet piles that occurs due to the difference in water levels on the two sides of these structures. In this section, we solve a seepage problem associated to a soil embankment and a sheet pile structure with the configuration shown in Figure 5.6.

Let Ω denote the soil embankment. Our goal is to find the head distribution $h(x,y)$ in soil and the discharge $\boldsymbol{q}$ through soil with

$$\boldsymbol{q} = -\boldsymbol{D}(h) = \begin{bmatrix} -\kappa_x \partial_x h \\ -\kappa_y \partial_y h \end{bmatrix},$$

where the parameters κ_x and κ_y are respectively the permeability coefficients in the x- and the y-directions. To state a suitable boundary value problem for this problem, we assume that $\partial\Omega = \Gamma_D \cup \Gamma_N$, where Γ_N denotes the impervious boundaries including the left, the bottom, and the right edges of soil embankment together with the edges around the sheet pile and $\Gamma_D = \Gamma_D^1 \cup \Gamma_D^2$, where Γ_D^1 and Γ_D^2 are the portions of the top boundary of the soil embankment located on the left and the right side of the sheet pile, respectively. Also suppose $\boldsymbol{n}$ is the unit normal vector field of $\partial\Omega$. Then, we

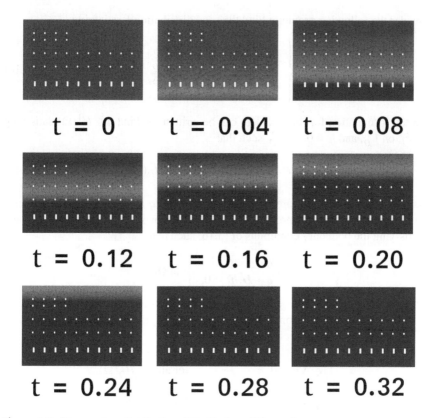

Figure 5.5 Temperature distribution of the blade at different time steps.

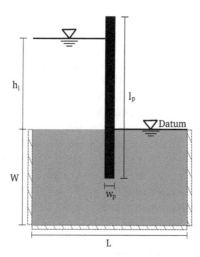

Figure 5.6 Configuration of the seepage problem.

consider the following problem: Find the head distribution h satisfying

$$\begin{cases} -\operatorname{div}\boldsymbol{D}(h) = 0, & \text{in } \Omega, \\ h = h_l, & \text{on } \Gamma_D^1, \\ h = 0, & \text{on } \Gamma_D^2, \\ \boldsymbol{q} \cdot \boldsymbol{n} = 0, & \text{on } \Gamma_N. \end{cases} \qquad (5.16)$$

A weak formulation for this problem can be written as: Find $h \in H^1(\Omega)$ satisfying $h = h_l$, on Γ_D^1, and $h = 0$, on Γ_D^2, such that

$$\int_\Omega \boldsymbol{D}(h) \cdot \nabla v = 0, \quad \text{for all } v \in H_D^1(\Omega). \qquad (5.17)$$

Since in addition to h, we are interested to compute $\boldsymbol{q}$ as well, we also consider a mixed formulation for this problem based on the following strong form: Find h and $\boldsymbol{q}$ satisfying the boundary conditions of (5.16) such that

$$\left.\begin{array}{l} \operatorname{div}\boldsymbol{q} = 0, \\ \boldsymbol{q} + \boldsymbol{D}(h) = 0, \end{array}\right\} \text{ in } \Omega.$$

Let V be either $[H^1(\Omega)]^2$ or $H(\operatorname{div};\Omega)$. One can write the following mixed formulation based on the above strong form: Find $\boldsymbol{q} \in V$ and $h \in H^1(\Omega)$ satisfying $h = h_l$, on Γ_D^1, and $h = 0$, on Γ_D^2, such that

$$\begin{aligned} \int_\Omega \boldsymbol{q} \cdot \nabla v &= 0, & \text{for all } v \in H_D^1(\Omega), \\ \int_\Omega \boldsymbol{q} \cdot \boldsymbol{w} + \int_\Omega \boldsymbol{D}(h) \cdot \boldsymbol{w} &= 0, & \text{for all } \boldsymbol{w} \in V. \end{aligned} \qquad (5.18)$$

The single-field problem (5.17) and the mixed problem (5.18) with $V = H(\operatorname{div};\Omega)$ are implemented in the function `seepage_solver`:

```
def seepage_solver(N, kx, ky, degreeCG, degreeDiv):
    # create mesh and define function space
    domain_vertices = [Point(0.0, 0.0),
                       Point(L, 0.0),
                       Point(L, W),
                       Point(L/2 + Wp/2, W),
                       Point(L/2 + Wp/2, W/2 + 1.0),
                       Point(L/2 - Wp/2, W/2 + 1.0),
                       Point(L/2 - Wp/2, W),
                       Point(0.0, W)]
    domain = Polygon(domain_vertices)
    mesh = generate_mesh(domain, N)

    # define mixed function spaces
    LG = FiniteElement("CG", mesh.ufl_cell(), degreeCG)
```

```
RT = FiniteElement("RT", mesh.ufl_cell(), degreeDiv)
V_mixed = FunctionSpace(mesh, MixedElement([LG, RT]))

# define single-field function spaces
V_single = FunctionSpace(mesh, "Lagrange", degreeCG)
Vec_h = VectorFunctionSpace(mesh, "Lagrange", degreeCG+1)

# define boundary conditions
def inflow(x, on_boundary):
    tol = 1E-12       # tolerance for coordinate comparisons
    return on_boundary and abs(x[1] - 10.0) < tol\
           and 0.0 < x[0] < 9.0
def outflow(x, on_boundary):
    tol = 1E-12       # tolerance for coordinate comparisons
    return on_boundary and abs(x[1] - 10.0) < tol\
           and 11.0 < x[0] < 20.0

# boundary function for mixed
bc_mixed = [DirichletBC(V_mixed.sub(0), Constant(h_l),\
                        inflow),
            DirichletBC(V_mixed.sub(0), Constant(0.0),\
                        outflow)]

# boundary function for single-field
bc_single = [DirichletBC(V_single, Constant(h_l),\
                         inflow),
             DirichletBC(V_single, Constant(0.0), \
                         outflow)]

# define the weak formulation for mixed
(h_mixed, q_mixed) = TrialFunctions(V_mixed)
(v_mixed, w_mixed) = TestFunctions(V_mixed)
f = Constant((0.0))                     # a dummy function
Dh_mixed = as_vector([kx*h_mixed.dx(0), ky*h_mixed.dx(1)])
LHS = ( inner(q_mixed, grad(v_mixed)) +
        inner(q_mixed, w_mixed) +
        inner(Dh_mixed, w_mixed))*dx
RHS = f*v_mixed*dx

# define the weak formulation for single-field
h_single = TrialFunction(V_single)
v_single = TestFunction(V_single)
Dh_single = as_vector([kx*h_single.dx(0),\
                       ky*h_single.dx(1)])
```

```
A = inner(Dh_single, grad(v_single))*dx
F = f*v_single*dx

# compute finite element solution for mixed
mixed_soln = Function(V_mixed)
solve(LHS == RHS, mixed_soln, bc_mixed)
(h_mixed, q_mixed) = mixed_soln.split()

# compute finite element solution for single-field
h_single = Function(V_single)
solve(A == F, h_single, bc_single)

# compute discharge by gradient
q_single = project(-grad(h_single), Vec_h)

# obtaining L2 norms
Norm_h_mixed = norm(h_mixed, norm_type="L2")
Norm_h_single = norm(h_single, norm_type="L2")
Norm_q_mixed = norm(q_mixed, norm_type="L2")
Norm_q_single = norm(q_single, norm_type="L2")
NORM = [Norm_h_mixed, Norm_h_single,\
        Norm_q_mixed, Norm_q_single]

return NORM, mesh.hmax(), q_mixed,\
       q_single, h_mixed, h_single
```

Notice that we used a projection approach similar to (4.56) to compute q in the single-field problem. In the following code, seepage_solver is employed to compute some results.

```
# mesh resolutions
divisions = [4, 8, 12, 16, 20, 24, 30]
L = 20.0                      # length of soil embankment
W = 10.0                      # width of soil embankment
h_l = 1.0                     # water head
Wp = 2.0                      # sheet pile width
kx = 0.00000037              # permeability in x-direction
ky = 0.00000042              # permeability in y-direction
degreeCG = 1 ; degreeDiv = 1 # degrees of CG and RT elements

h_m = []   # element sizes
E = []   # norms
Norm_h_mixed = np.zeros((len(divisions),))   # mixed-head
Norm_h_single = np.zeros((len(divisions),)) # single-head
```

```
Norm_q_mixed = np.zeros((len(divisions),))  # mixed-discharge
Norm_q_single = np.zeros((len(divisions),)) # single-discharge

for (i, N) in enumerate(divisions):
    # solver
    Tnorm, hmesh, q_mixed, q_single, h_mixed, h_single\
    = seepage_solver(N, kx, ky, degreeCG, degreeDiv)

    h_m.append(hmesh)
    E.append(Tnorm)
    Norm_h_mixed[i] = norm(h_mixed, norm_type="L2")
    Norm_h_single[i] = norm(h_single, norm_type="L2")
    Norm_q_mixed[i] = norm(q_mixed, norm_type="L2")
    Norm_q_single[i] = norm(q_single, norm_type="L2")

for i in range(0, len(E)):
    print('h=%.3f | head_mixed_norm=%8.4E\
           | head_single_norm=%8.4E |flux_mixed_norm=%8.4E\
           | flux_single_norm=%8.4E'
    % (h_m[i], E[i][0], E[i][1], E[i][2], E[i][3]))

file = File("Ch5_seepage/head_mixed.pvd")
file << h_mixed
file = File("Ch5_seepage/discharge_mixed.pvd")
file << q_mixed
file = File("Ch5_seepage/head_single.pvd")
file << h_single
file = File("Ch5_seepage/discharge_single.pvd")
file << q_single
```

Figure 5.7 shows L^2-norms of the head distribution and discharge computed by the single-field and the mixed methods by using the parameters given in the above code. As can be expected, the mixed formulation yields a more accurate approximation of discharge. These head distribution and discharge are plotted on the soil embankment configuration in Figure 5.8.

5.7   SOIL CONSOLIDATION

Consolidation problems describe gradual volume changes in saturated soils due to the pore water dissipation under mechanical loads. We solve a 2D consolidation problem associated with the clay embankment Ω shown in Figure 5.9, which consists of two layers Ω_0 and Ω_1 of saturated clay. Let Γ_{top} denote the top boundary of Ω.

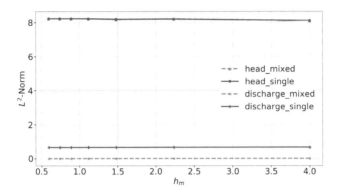

Figure 5.7 L^2-norms of the head distribution and discharge versus the mesh size h_m computed by using the single-field formulation (5.17) and the mixed formulation (5.18). Finite elements of degree 1 are used for these results.

Figure 5.8 The head distribution and discharge streams through soil embankment computed based on the single field formulation (5.17) (the right panel) and the mixed formulation (5.18) (the left panel). Colors denote the head distribution. Finite elements of degree 1 are used for these results.

We seek for a time-dependent pore water pressure $u(x,y,t)$ such that

$$\begin{cases} \partial_t u = c_v \Delta u, & \text{in } \Omega, \ t \in [0,T], \\ \partial_y u = Q, & \text{on } \Gamma_{\text{top}}, \ t \in [0,T], \\ \boldsymbol{n} \cdot \nabla u = 0, & \text{on } \partial\Omega - \Gamma_{\text{top}}, \ t \in [0,T], \\ u = u_0, & \text{at } t = 0, \end{cases}$$

where $\boldsymbol{n}$ is the unit normal vector field of $\partial\Omega$ and

$$c_v = \frac{\kappa m_v}{\gamma_w},$$

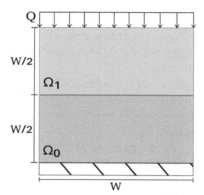

Figure 5.9 Configuration of a two-layers clay embankment with a drained edge on its bottom boundary. The uniform load Q is applied on the top boundary.

is the coefficient of consolidation, which is determined by the permeability coefficient κ, the compressibility coefficient m_v, and the specific weight of water γ_w. In this problem, we assume the layers have different permeability and compressibility coefficients.

A weak formulation for the above initial-boundary value problem reads: Find $u(x,y,t)$ with $u(\cdot,t) \in H^1(\Omega)$ for any $t \in [0,T]$ such that

$$\begin{cases} \int_\Omega v \partial_t u + \int_\Omega c_v \nabla u \cdot \nabla v = \int_{\Gamma_{top}} c_v Q v, & \text{for all } v \in H^1(\Omega), \\ u = u_0, & \text{at } t = 0. \end{cases} \qquad (5.19)$$

By using an H^1-conformal finite element space V_h and the implicit Euler method with a uniform time increment $S = t_{i+1} - t_i$, we can discretize (5.19) as: Given $u_h^i \in V_h$, find $u_h^{i+1} \in V_h$ such that

$$\int_\Omega u_h^{i+1} v_h + S \int_\Omega c_v \nabla u_h^{i+1} \cdot \nabla v_h = S \int_{\Gamma_{top}} c_v Q v_h + \int_\Omega u_h^i v_h, \text{ for all } v_h \in V_h.$$

This consolidation problem is implemented in the following program.

```
# parameters
T = 431100             # final time
num_steps = 2000       # number of time steps
S = T / num_steps      # time step size
N = 40                 # mesh resolution
degreeCG = 1           # degree of polynomial
W = 10.0               # length and width
gamma_w = 9.81         # water specific weigth
k_0 = 0.0022           # permeability of clay0
k_1 = 0.037            # permeability of clay1
mv_0 = 0.09            # compressibility of clay0
mv_1 = 0.03            # compressibility of clay1
```

```
# create mesh and define function space
domain = Rectangle(Point(0.0, 0.0), Point(W, W))
mesh = generate_mesh(domain, N)
V = FunctionSpace(mesh, 'CG', degreeCG)

# define subdomains
class Omega0(SubDomain):
    def inside(self, x, on_boundary):
        return True if x[1] <= W/2 else False

class Omega1(SubDomain):
    def inside(self, x, on_boundary):
        return True if x[1] >= W/2 else False
subdomains = MeshFunction('size_t', mesh\
                          , mesh.topology().dim())
boundary_parts = MeshFunction('size_t', mesh\
                          , mesh.topology().dim()-1)

# mark subdomains
subdomain0 = Omega0()
subdomain0.mark(subdomains, 0)
subdomain1 = Omega1()
subdomain1.mark(subdomains, 1)

# create constant function in each subdomains
V0 = FunctionSpace(mesh, 'DG', 0)
k  = Function(V0)
mv = Function(V0)

# extracting the corresponding subdomain number of a cell
# and assign the corresponding k and mv values
k_values = [k_0, k_1]     # values of k in the two subdomains
mv_values = [mv_0, mv_1]  # values of mv in the two subdomains
browse = np.asarray(subdomains.array(), dtype=np.int32)
k.vector()[:] = np.choose(browse, k_values)
mv.vector()[:] = np.choose(browse, mv_values)

# mark top boundary
class TopBoundary(SubDomain):
    def inside(self, x, on_boundary):
        tol = 1E-14   # tolerance for coordinate comparisons
        return on_boundary and abs(x[1] - W) < tol
```

```
Gamma_1 = TopBoundary()
Gamma_1.mark(boundary_parts, 2)

ds = Measure("ds", domain=mesh, \
             subdomain_data=boundary_parts)

# define initial value
u1 = Constant((50.0))
u_0 = interpolate(u1, V)

# define variational problem
u = TrialFunction(V)
v = TestFunction(V)
Q = Constant((-50.0))                # loading
cv = k/gamma_w*mv                    # consolidation parameter
a = u*v*dx + cv*S*dot(grad(u), grad(v))*dx
L = u_0*v*dx + cv*S*dot(Q,v)*ds(2)

# define time-stepping vectors
time = np.linspace(0, T, num_steps+1)

# produce a vtkfile for visualization
vtkfile = File("Ch5_consolidation/PorePressure.pvd")

# time-stepping
t = 0
while t <= T:
    u = Function(V)

    # compute solution
    solve(a == L, u)

    # save solution to vtk format
    vtkfile << (u,t)

    # update previous solution
    u_0.assign(u)

    # update current time
    t += S
```

Notice that MeshFunction is used for specifying the top boundary and also for defining different properties of the layers. Figure 5.10 depicts pore water distributions at different time steps computed by using the parameter values given in the above program.

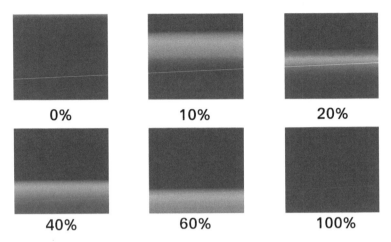

Figure 5.10 Pore water pressure distributions at some time steps with the specified degrees of settlement.

5.8 THE STOKES EQUATION FOR INCOMPRESSIBLE FLUIDS

The Stokes equation is a simple model for studying motions of fluids. Here we solve this equation to model steady flows of incompressible fluids, that is, fluids with a constant density.

Let Ω denote a 2D or a 3D domain occupied by a fluid. We seek for the velocity $\boldsymbol{v} = (v_1, \ldots, v_n)$, $n = 2$ or 3, and the pressure p of the fluid satisfying the boundary value problem

$$\begin{cases} -\mu \Delta \boldsymbol{v} + \nabla p = \boldsymbol{f}, & \text{in } \Omega, \\ \operatorname{div} \boldsymbol{v} = 0, & \text{in } \Omega, \\ \boldsymbol{v} = \boldsymbol{v}_0, & \text{on } \partial\Omega, \end{cases} \tag{5.20}$$

where the dynamic viscosity μ is a constant, $\boldsymbol{f}$ is a given body force, $\boldsymbol{v}_0$ is the velocity of the fluid at the boundary, and the vector Laplacian $\Delta \boldsymbol{v}$ is defined componentwise, that is, $\Delta \boldsymbol{v} = (\Delta v_1, \ldots, \Delta v_n)$. We only consider Dirichlet boundary condition in this section. Other choices of boundary conditions for the Stokes equation are discussed in [6, Section 4.1.4]; Also see Exercise 5.13.

To obtain a weak formulation for (5.20), let q be an arbitrary real-valued function and let $\boldsymbol{w} = (w_1, \ldots, w_n)$ be an arbitrary vector field belonging to $V = [H_0^1(\Omega)]^n$, that is, the components of $\boldsymbol{w}$ belong to $H^1(\Omega)$ and vanish on the boundary. Multiplying the second equation of (5.20) by q and taking the inner product of the first equation with $\boldsymbol{w}$ and applying Green's formulas (2.4) and (2.5) result in the following weak formulation for (5.20): Find $\boldsymbol{v} \in [H^1(\Omega)]^n$ and $p \in P$ such that $\boldsymbol{v} = \boldsymbol{v}_0$ on $\partial\Omega$ and

$$\begin{aligned} \int_\Omega \mu \nabla \boldsymbol{v} : \nabla \boldsymbol{w} - \int_\Omega p \operatorname{div} \boldsymbol{w} &= \int_\Omega \boldsymbol{f} \cdot \boldsymbol{w}, && \text{for all } \boldsymbol{w} \in V, \\ -\int_\Omega q \operatorname{div} \boldsymbol{v} &= 0, && \text{for all } q \in P, \end{aligned} \tag{5.21}$$

where

$$\nabla v : \nabla w = \sum_{i=1}^{n} \nabla v_i \cdot \nabla w_i = \sum_{i,j=1}^{n} \partial_{x_j} v_i \, \partial_{x_j} w_i. \tag{5.22}$$

The pressure p in (5.21) can belong to $L^2(\Omega)$ or, if p is smooth enough, $H^1(\Omega)$; See the discussion of Section 2.8 regarding Green's formulas in terms of Sobolev spaces. However, the above problem will not have a unique solution if we let the space P to be either of these spaces. This follows from the fact that if p is a solution of (5.20), then so is $p + c$, where c is any arbitrary constant.

To ensure that (5.21) is well-posed, we have to remove the above degree of freedom for pressure. This can be achieved by imposing additional conditions on p. One approach is to require that $\int_\Omega p = 0$, and seek for p in the space of L^2- or H^1-functions with zero average. Another approach is to fix the value of pressure at one point of the domain. The first approach is more suitable for theoretical analysis while the second one is more convenient for computations. Later, we will employ the second approach to implement (5.21) in FEniCS.

It turns out that the solution (v, p) of the weak formulation (5.21) is a saddle-point of the functional

$$J(w, q) = \frac{1}{2} \int_\Omega \mu \nabla w : \nabla w - \int_\Omega q \operatorname{div} w + \int_\Omega f \cdot w.$$

For this reason, the weak formulation (5.21) is also called *a mixed formulation* for the boundary value problem (5.20). This interpretation of a mixed formulation is different from the one mentioned in Sections 4.8.1 and 5.2.

Finite element methods based on the mixed formulation (5.21) are called mixed finite element methods for the boundary value problem (5.20). Similar to the other examples of mixed finite element methods we discussed so far, well-posedness of the mixed formulation (5.21) does not necessarily implies well-posedness of the associated mixed finite element methods. This can be explained by using the stiffness matrix of these finite element methods as follows.

Suppose $V_h \subset V$ and $P_h \subset P$ are finite element spaces with the global shape functions $\{\phi_i\}_{i=1}^{r}$ and $\{\psi_i\}_{i=1}^{s}$, respectively. The stiffness matrix of a conformal mixed finite element method based on (5.21) reads

$$\begin{bmatrix} \mathbf{A}_{r \times r} & \mathbf{B}_{r \times s}^T \\ \hline \mathbf{B}_{s \times r} & \mathbf{0} \end{bmatrix},$$

with

$$\mathbf{A}_{ij} = \int_\Omega \mu \nabla \phi_i : \nabla \phi_j, \quad \mathbf{B}_{ij} = -\int_\Omega \psi_i \operatorname{div} \phi_j.$$

Following the discussion of Section 4.8.2, we conclude that if this mixed finite element method is well-posed then $\mathbf{B}$ should be full rank. As a consequence of the rank-nullity theorem, $\mathbf{B}$ is not full rank if $r < s$, that is, if $\dim V_h < \dim P_h$. More generally, a necessary and sufficient condition for $\mathbf{B}$ to be full rank and for the associated finite element methods to be well-posed is the existence of a constant $\alpha_h > 0$ such that

$$\inf_{q_h \in P_h} \sup_{\boldsymbol{v}_h \in V_h} \frac{\int_\Omega q_h \, \mathrm{div} \, \boldsymbol{v}_h}{\|q_h\|_2 \, \|\boldsymbol{v}_h\|_{1,2}} \geq \alpha_h. \tag{5.23}$$

In the literature, the above inf-sup condition is called *the Babuška-Brezzi condition*.

Example 5.4. The choice of simplex of type (1) for both velocity and pressure leads to unstable mixed methods as it violates (5.23). For any integer $k \geq 2$, the choice of simplex of type (k) for velocity and simplex of type $(k-1)$ for pressure leads to stable mixed method. This stable choice is called *the Taylor-Hood element* in the literature. A detailed analysis of mixed finite element methods for the Stokes equation is available in [6, Chapter 4] and [4, Chapter 8].

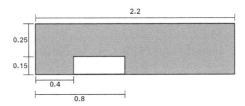

Figure 5.11   The geometry of the problem of flows over a rectangular step.

To implement the boundary value problem (5.20) in FEniCS, we consider a 2D flow over a rectangular step with the geometry shown in Figure 5.11. We assume that $\boldsymbol{f} = 0$, and that the boundary velocity is zero on the top and the bottom boundaries of the domain. On the left and the right boundaries of the domain, we impose the parabolic velocity profile $\boldsymbol{v}(y) = \left(\frac{4U}{H^2}y(H-y), 0\right)$, where H is the height of the domain and U is the maximum velocity. This problem can be implemented using FEniCS as follows.

```
# parameters
n = 330                 # number of divisions
U = 15.0                # maximum velocity
mu = 1.0                # kinematic viscosity
k = 2                   # degree of polynomial


# generate mesh
step = Rectangle(Point(0.4, 0.0), Point(0.8, 0.15))
Rec = Rectangle(Point(0.0, 0.0), Point(2.2, 0.4))
geometry = Rec - step
mesh = generate_mesh(geometry, n)


# define function spaces
T = VectorElement("CG", mesh.ufl_cell(), k)
```

```
Q = FiniteElement("CG", mesh.ufl_cell(), k-1)
mixed = T * Q
W = FunctionSpace(mesh,mixed)

# define subdomains
inflow = 'near(x[0], 0)'
outflow = 'near(x[0], 2.2)'
topnbottom = 'near(x[1], 0) || near(x[1], 0.4)'
small_rec = """on_boundary && x[0] > 0.4001 && \
               x[0] < 0.8001 && x[1] < 0.15001"""

# pinpoint for removing the degree of freedom of pressure
class PinPoint(SubDomain):
    def inside(self, x, on_boundary):
        return x[0] < DOLFIN_EPS and x[1] < DOLFIN_EPS
pinpoint = PinPoint()

# no-slip boundary condition for velocity at top and bottom
noslip = Constant((0.0, 0.0))
bc0 = DirichletBC(W.sub(0), noslip, topnbottom)

# no-slip boundary condition for velocity at step
bc1 = DirichletBC(W.sub(0), noslip, small_rec)

# inflow/outflow on left and right boundary
velocity_profile = \
  Expression((('4.0*U*x[1]*(0.4 - x[1]) / pow(0.4, 2)','0.0'),
              U = U, degree=4 + k)
bc2 = DirichletBC(W.sub(0), velocity_profile, inflow)
bc3 = DirichletBC(W.sub(0), velocity_profile, outflow)

# pressure at pinpoint
zero = Constant((0.0))
bc4 = DirichletBC(W.sub(1), zero, pinpoint, "pointwise")

# all boundary conditions
bcs = [bc0, bc1, bc2, bc3, bc4]

# variational problem
(v, p) = TrialFunctions(W)
(w, q) = TestFunctions(W)
f = Constant((0.0, 0.0))
a = mu*inner(grad(v), grad(w))*dx - div(w) *p*dx - q*div(v)*dx
L = inner(f, w)*dx
```

```
# solver
V = Function(W)
solve(a == L, V, bcs)

# extracting sub-functions
v, p = V.split()

# saving solution in VTK format
ufile_pvd = File("Ch5_Stokes/velocity.pvd")
ufile_pvd << v
pfile_pvd = File("Ch5_Stokes/pressure.pvd")
pfile_pvd << p
```

In this code, to remedy the nonuniqueness of pressure, the value of pressure is kept to be zero at the origin by using the class PinPoint. Figure 5.12 shows the velocity and the pressure distributions computed by using the parameters given in the above code and the Taylor-Hood element with degree $k = 2$ for velocity and degree $k = 1$ for pressure.

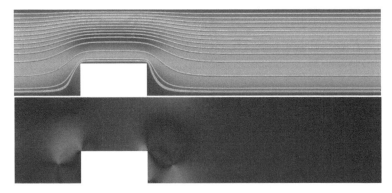

Figure 5.12 A steady fluid flow over a step: (top) The speed distribution and velocity stream lines; (bottom) The pressure distribution.

5.9 LINEARIZED ELASTICITY

Let Ω denote an n-dimensional linearly elastic body, $n = 2, 3$. The equilibrium equation of this body under a body force f reads

$$\mathrm{div}\,\boldsymbol{\sigma} = -f, \text{ in } \Omega,$$

where $\boldsymbol{\sigma}$ is the stress tensor with the components σ_{ij}, and the i-th component of the vector field $\mathrm{div}\,\boldsymbol{\sigma}$ is given by $\sum_{j=1}^{n} \partial_{x_j} \sigma_{ij}$. The constitutive equation of isotropic, linearly elastic bodies in terms of the linear strain tensor $\boldsymbol{\varepsilon}$ can be expressed as

$$\boldsymbol{\sigma}(\boldsymbol{\varepsilon}) = \lambda \operatorname{tr}(\boldsymbol{\varepsilon})\boldsymbol{I} + 2\mu\boldsymbol{\varepsilon},$$

where $\lambda, \mu > 0$ are the Lame parameters, and $\boldsymbol{I}$ is the identity tensor with its components being the Kronecker delta δ_{ij}. The Lame parameters can be expressed in terms of the Young modulus E and the Poisson ratio v by using the relations

$$\lambda = \frac{Ev}{(1+v)(1-2v)}, \quad \mu = \frac{E}{2(1+v)}.$$

In components, the constitutive equation states that $\sigma_{ij} = \lambda\left(\sum_{k=1}^{n} \varepsilon_{kk}\right)\delta_{ij} + 2\mu\,\varepsilon_{ij}$. The linear strain tensor is defined in terms of the displacement vector $\boldsymbol{u}$ as

$$\boldsymbol{\varepsilon}(\boldsymbol{u}) = \frac{1}{2}\left(\nabla\boldsymbol{u} + (\nabla\boldsymbol{u})^{T}\right), \tag{5.24}$$

or equivalently, $\varepsilon_{ij} = \frac{1}{2}(\partial_{x_j}u_i + \partial_{x_i}u_j)$. By using (5.24), the constitutive equation may be stated in terms of $\boldsymbol{u}$ as

$$\boldsymbol{\sigma}(\boldsymbol{u}) = \lambda(\operatorname{div}\boldsymbol{u})\boldsymbol{I} + \mu\left(\nabla\boldsymbol{u} + (\nabla\boldsymbol{u})^{T}\right). \tag{5.25}$$

In linearized elasticity, the displacement is usually considered as the main unknown. This leads to the following boundary value problem for linearized elasticity: Suppose $\partial\Omega = \Gamma_D \cup \Gamma_N$, and let $\boldsymbol{n}$ be the outward unit normal vector field of $\partial\Omega$. Then, we look for the displacement field $\boldsymbol{u} = (u_1, \ldots, u_n)$ such that

$$\begin{cases} \operatorname{div}\boldsymbol{\sigma}(\boldsymbol{u}) = -\boldsymbol{f}, & \text{in } \Omega, \\ \boldsymbol{u} = \boldsymbol{u}_0, & \text{in } \Gamma_D, \\ \boldsymbol{\sigma}\boldsymbol{n} = \boldsymbol{T}, & \text{on } \Gamma_N. \end{cases}$$

The vector field $\boldsymbol{T}$ is called the traction vector field and it represents external surface loads imposed at the boundary. In components, the boundary condition on Γ_N reads $\sum_j \sigma_{ij}n_j = T_i$.

By taking the inner product of the equilibrium equation with an arbitrary H^1 vector field $\boldsymbol{v}$ that vanishes on Γ_D, that is, $\boldsymbol{v} \in [H_D^1(\Omega)]^n$, and applying Green's formula, we obtain the following weak formulation for linearized elasticity: Find $\boldsymbol{u} \in [H^1(\Omega)]^n$ with $\boldsymbol{u} = \boldsymbol{u}_0$, on Γ_D, such that

$$\int_{\Omega} \boldsymbol{\sigma}(\boldsymbol{u}) : \nabla\boldsymbol{v} = \int_{\Omega} \boldsymbol{f}\cdot\boldsymbol{v} + \int_{\Gamma_N} \boldsymbol{T}\cdot\boldsymbol{v}, \quad \text{for all } \boldsymbol{v} \in [H_D^1(\Omega)]^n, \tag{5.26}$$

where $\boldsymbol{\sigma} : \nabla\boldsymbol{v} = \sum_{i,j=1}^{n} \sigma_{ij}\,\partial_{x_j}v_i$.

One can obtain conformal finite element methods for linearized elasticity by using Lagrange elements. As an example, we compute the deflection of the clamped beam of Figure 5.2 by considering the beam as a 3D body. More specifically, let the cross-section of the beam be the T-section shown in Figure 5.13 and let the distributed load $\boldsymbol{f}$, the bending moment M, and the body force vanish. Also suppose the right end is under a constant shear load $\boldsymbol{T} = \boldsymbol{\sigma}\boldsymbol{n}$. The following code computes the resulting displacement of the beam.

```
# beam parameters
L = 20; W = 6                          # length and width of the beam
mu = 80.194; lambda_ = 400889.8  # material properties
k = 2                                  # degree of element
N = 50                                 # number of divisions


# create Tbar geometry and generate mesh
Box1 = Box(Point(0,0,0), Point(W,W,L))
Box2 = Box(Point(0,0,0), Point(W/3,2*W/3,L))
Box3 = Box(Point(2*W/3,0,0), Point(W,2*W/3,L))
TBox = Box1-Box2-Box3
mesh = generate_mesh(TBox, N)


# define function space
V = VectorFunctionSpace(mesh, 'P', k)


# define boundary conditions
boundary_parts = MeshFunction("size_t",
                                mesh, mesh.topology().dim()-1)


# mark boundary facet with Z=L as subdomain 3
class RightBoundary(SubDomain):
    def inside(self, x, on_boundary):
        tol = 1E-12    # tolerance for coordinate comparisons
        return on_boundary and abs(x[2] - L) < tol
Gamma_R = RightBoundary()
Gamma_R.mark(boundary_parts, 3)


# boundary function for the face Z=0
def left_f(x, on_boundary):
    tol = 1E-12    # tolerance for coordinate comparisons
    return on_boundary and abs(x[2]) < tol


bc = DirichletBC(V, Constant((0, 0, 0)), left_f)


# define strain and stress
def epsilon(u):
    return 0.5*(nabla_grad(u) + nabla_grad(u).T)


def sigma(u):
    return lambda_*nabla_div(u)*Identity(d) + 2*mu*epsilon(u)


ds = Measure("ds", domain=mesh,
             subdomain_data=boundary_parts) # boundary integral
```

```
# define variational problem
u = TrialFunction(V)
v = TestFunction(V)
T = Constant((0, -3.0, 0))
d = u.geometric_dimension()  # space dimension
a = inner(sigma(u), epsilon(v))*dx
L = dot(T, v)*ds(3)

# compute solution
u = Function(V)
solve(a == L, u, bc)

# saving solution in VTK format
ufile_pvd = File("Ch5_LinElas/deformation.pvd")
ufile_pvd << u
```

A deformed configuration of the beam computed by using the parameter values given in the above code is shown in Figure 5.13. Colors in the deformed configuration indicate the distribution of the norm of the displacement field.

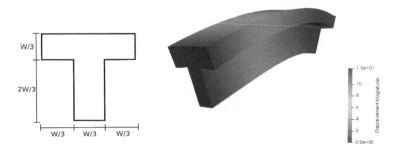

Figure 5.13   The cross-section and the deformed configuration of the 3D beam example.

5.10 LINEARIZED ELASTODYNAMICS: THE HAMBURG WHEEL-TRACK TEST

Hamburg Wheel-Track Test (HWTT) AASHTO T 324 is a device that uses a loaded wheel to apply a moving load on compacted asphalt mixture specimens to simulate traffic cycles on asphalt pavements. Figure 5.14 schematically depicts a HWTT specimen. The diameter, the width, and the mass of the wheel on the top of the specimen are 8 in (0.203 m), 1.85 in (0.047 m), and 160 lb (72.57 kg), respectively. The wheel moves back and forth on the top of the specimen with the frequency 52 ± 2 pass/minute. The shaded strip in Figure 5.14 shows the path of the wheel.

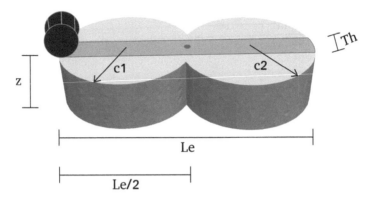

Figure 5.14 Geometry of a HWTT specimen: The shaded rectangle strip of the length Le and the width Th on top of the specimen shows the path of the wheel as it periodically moves between the left and the right sides of the specimen.

Assuming responses of the specimen Ω can be modeled by using linearized elasticity, we can express the initial-boundary value problem for approximating the time-dependent displacement field $\boldsymbol{u}$ of the specimen as follows: Let $\partial\Omega = \Gamma_D \cup \Gamma_N$, where Γ_D denotes the bottom and side boundaries and Γ_N is the top boundary of Ω. Then, we look for the vector field $\boldsymbol{u}$ such that

$$\begin{cases} \operatorname{div}\boldsymbol{\sigma}(\boldsymbol{u}) + \boldsymbol{f} = \rho\partial_{tt}\boldsymbol{u}, & \text{in } \Omega,\ t \in [0,T], \\ \boldsymbol{u} = 0, & \text{on } \Gamma_D,\ t \in [0,T], \\ \boldsymbol{\sigma}\boldsymbol{n} = \boldsymbol{T}(t), & \text{on } \Gamma_N,\ t \in [0,T], \\ \boldsymbol{u} = \partial_t\boldsymbol{u} = 0, & \text{at } t = 0, \end{cases}$$

where $\boldsymbol{f}$ is the body force, ρ is the density of the specimen, and the constitutive equation $\boldsymbol{\sigma}(\boldsymbol{u})$ is given in (5.25). In this problem, the traction vector field is time-dependent.

A weak form for the above problem reads: Find $\boldsymbol{u}(x,y,z,t)$ with $\boldsymbol{u}(\cdot,t) \in [H_D^1(\Omega)]^3$ such that

$$\begin{cases} \int_\Omega \rho\partial_{tt}\boldsymbol{u}\cdot\boldsymbol{v} + \int_\Omega \boldsymbol{\sigma}(\boldsymbol{u}):\nabla\boldsymbol{v} = \int_\Omega \boldsymbol{f}\cdot\boldsymbol{v} + \int_{\Gamma_N}\boldsymbol{T}\cdot\boldsymbol{v}, & \text{for all } \boldsymbol{v}\in[H_D^1(\Omega)]^3, \\ \boldsymbol{u} = \partial_t\boldsymbol{u} = 0, & \text{at } t = 0. \end{cases}$$

The three-point centered-difference formula of Exercise 4.19 allows us to write the following implicit finite element method for the above weak problem: Let $V_h \subset [H_D^1(\Omega)]^3$ be a finite element space induced by 3-simplex of type (k) and let $S = t_{i+1} - t_i$, be a uniform time increment. Then, given $\boldsymbol{u}_h^i, \boldsymbol{u}_h^{i+1} \in V_h$, find $\boldsymbol{u}_h^{i+2} \in V_h$ such that

$$3\int_\Omega \boldsymbol{u}_h^{i+2}\boldsymbol{v}_h + S^2\int_\Omega \boldsymbol{\sigma}(\boldsymbol{u}^{i+2}):\nabla\boldsymbol{v}_h =$$
$$2\int_\Omega \boldsymbol{u}_h^{i+1}\boldsymbol{v}_h - \int_\Omega \boldsymbol{u}_h^i\boldsymbol{v}_h + S^2\int_\Omega \boldsymbol{f}\cdot\boldsymbol{v}_h + S^2\int_{\Gamma_N}\boldsymbol{T}^{i+2}\cdot\boldsymbol{v}_h, \text{ for all } \boldsymbol{v}_h\in V_h,$$

(5.27)

with $\boldsymbol{u}_h^0 = \boldsymbol{u}_h^1 = 0$, and $\boldsymbol{T}^{i+2} = \boldsymbol{T}(t_{i+2})$.

To implement this finite element method, we assume the contact surface of the wheel and the specimen is a moving rectangle $Th \times Wc$, where Th is the width of the wheel path shown in Figure 5.14. The downward traction vector is induced by the wheel weight W_{wh} and its norm is $\frac{W_{wh}}{Th \times Wc}$. We assume the distance between the contact surface and the top left point of the specimen is given by the periodic function

$$g(t) = \frac{Le}{2}(1 - \cos \omega t),$$

where Le is the length of the specimen shown in Figure 5.14 and ω is the angular frequency of the motion. The method (5.27) can be implemented in FEniCS as follows.

```
# parameters
N = 50                        # mesh resolution
px1 = 0.075                   # center of first cylinder
px2 = 0.212                   # center of second cylinder
c1 = c2 = 0.075              # cylinders radius
Th = 0.04699                  # rectangle strip width
Le = 0.287                    # strip length
z = 0.06                      # cylinder height
omega = 2.618                 # frequency
W_wh = 47743.4211             # wheel weight
Wc = 0.04241                  # effective length
# time-stepping parameters
t = 0.0                       # initial time
T = 4.8                       # final time
Nsteps  = 960
S = Constant(T/Nsteps)

# elastic parameters
E   = 611829727.78674
nu = 0.35
mu      = Constant(E / (2.0*(1.0 + nu)))
lmbda = Constant(E*nu / ((1.0 + nu)*(1.0 - 2.0*nu)))

# mass density
rho = Constant(2700)          # asphalt mixture density

# create HWTT sample geometry and generate mesh
cylinder1 = Cylinder(Point(px1, px1, 0.0), \
                     Point(px1, px1, z), c1, c1)
cylinder2 = Cylinder(Point(px2, px1, 0.0), \
                     Point(px2, px1, z), c2, c2)
mold = cylinder1 + cylinder2
mesh = generate_mesh(mold, N)
```

```
# return symbolic physical coordinates for mesh.
x = SpatialCoordinate(mesh)

# define geometries to impose bcs
def cyl1(x, on_boundary):
    return on_boundary and np.isclose(np.sqrt((x[0]-px1)**2 \
                    +(x[1]-px1)**2), 0.075, rtol=3e-2)
def cyl2(x, on_boundary):
    return on_boundary and np.isclose(np.sqrt((x[0]-px2)**2 \
                    +(x[1]-px1)**2), 0.075, rtol=3e-2)
def top(x, on_boundary):
    tol = 1E-12    # tolerance for coordinate comparisons
    return on_boundary and abs(x[2] - z) < tol
def bottom(x, on_boundary):
    tol = 1E-12    # tolerance for coordinate comparisons
    return on_boundary and abs(x[2]) < tol

# define function space for displacement,
# velocity and acceleration
V = VectorFunctionSpace(mesh, "CG", 1)

# test and trial functions
u = TrialFunction(V)
v = TestFunction(V)

# user-defined expression to impose traction on strip
class IC(UserExpression):
    def __init__(self, t, c1, Th, Le, omega, z, \
                W_wh, Wc, **kwargs):
        super().__init__(**kwargs)
        self.t = t
        self.omega = omega
        self.Le = Le
    def eval(self, value, x):
        value[0] = 0.0
        value[1] = 0.0
        gt = (self.Le/2)*(1 - np.cos(self.omega*self.t))
        if x[2] == z and  c1-(Th/2) <= x[1] <= c1+(Th/2)\
        and gt <= x[0] <= 1.04241*gt:
            value[2] = -W_wh/(Th*Wc)
        else :
            value[2] = 0.0
    def value_shape(self):
```

```
        return (3,)
Tr = IC(t, c1, Th, Le, omega, z, W_wh, Wc)


# current (unknown) displacement
uh = Function(V, name="Displacement")


# displacement for first and second iterations
u1= interpolate(Constant((0.0, 0.0, 0.0)), V)
u0= interpolate(Constant((0.0, 0.0, 0.0)), V)


# create mesh function over the cell facets
boundary_subdomains = MeshFunction("size_t", \
                   mesh, mesh.topology().dim() - 1)
boundary_subdomains.set_all(0)
mark_boundary = AutoSubDomain(top)
mark_boundary.mark(boundary_subdomains, 3)


# define measure for boundary condition integral
ds = Measure("ds", domain=mesh,
             subdomain_data=boundary_subdomains)


# set up boundary condition at bottom and sides
zero = Constant((0.0, 0.0, 0.0))
bc0 = DirichletBC(V, zero, bottom)
bc1 = DirichletBC(V, zero, cyl1)
bc2 = DirichletBC(V, zero, cyl2)
bcs = [bc0, bc1, bc2]


# define strain and stress
def epsilon(r):
    return 0.5*(nabla_grad(r) + nabla_grad(r).T)
def sigma(q):
    return lmbda*nabla_div(q)*Identity(len(q)) \
    + 2*mu*epsilon(q)


# residual
f = Constant((0, 0, 0))   # body force
a = rho*dot(u, v)*dx + S*S*inner(sigma(u), epsilon(v))*dx
L = S*S*rho*dot(f,v)*dx + S*S*dot(Tr,v)*ds(3) + \
    rho*dot(u0, v)*dx -2*rho*dot(u1, v)*dx


# time-stepping
time = np.linspace(0, T, Nsteps+1)
u_mid = np.zeros((Nsteps+1,)) # red dot on strip
```

```
# produce a vtkfile for visualization
vtkfile = File("Ch5_WheelTrack/rutting.pvd")

for (i, S) in enumerate(np.diff(time)):

    # solve
    A, b = assemble_system(a, L, bcs)
    solve(A, uh.vector(), b)

    # save solution to vtk format
    vtkfile << (uh,t)

    # update old fields with new quantities
    u0.assign(u1)
    u1.assign(uh)

    # update user-defined expression
    Tr.t = t

    # record midpoint displacement in z direction
    u_mid[i+1] = uh(Le/2, c1, z)[2]
    t = time[i+1]

    # print midpoint displacement and time while running
    print("midpoint disp %2.9f at Time %f " % (u_mid[i], t))
```

Figure 5.15 shows the displacement of the midpoint of the specimen, shown with a dark spot in Figure 5.14, versus time and Figure 5.16 shows rut configurations at some time steps. These results are computed by using the parameters given in the above program.

5.11 NONLINEAR ELASTICITY

Although studying nonlinear PDEs is out of the scope of this book, here we consider Newtons' iterations for solving a mixed finite element method for nonlinear elasticity. At each iteration, one solves a linear problem to obtain the approximate solution for the next iteration. Iterations are continued until the difference between approximate solutions of two consecutive iterations are sufficiently small. The main reason for discussing this example is to provide an application of the Nédélec finite element introduced in Section 3.2.4. For more details on the topic of this section, the reader is referred to [2].

The discussion of this section holds in 2D and 3D, however, we only consider the 2D case for the sake of simplicity. Suppose Ω denotes the reference configuration of a nonlinearly elastic body and let n be the outward unit normal vector field of

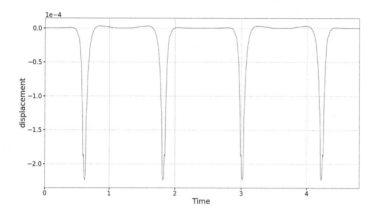

Figure 5.15 Displacement of the midpoint of the specimen versus time.

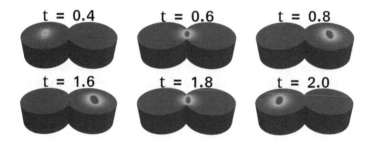

Figure 5.16 Rut configurations at different time steps.

$\partial \Omega = \Gamma_D \cup \Gamma_N$. We seek the displacement field $\boldsymbol{U}$, the displacement gradient $\boldsymbol{K}$, and the first Piola-Kirchhoff stress tensor $\boldsymbol{P}$ that satisfy

$$
\left.
\begin{aligned}
\operatorname{div} \boldsymbol{P} &= -\boldsymbol{f}, \\
\boldsymbol{K} - \nabla \boldsymbol{U} &= 0, \\
\boldsymbol{P} - \mathbb{P}(\boldsymbol{K}) &= 0,
\end{aligned}
\right\} \text{ in } \Omega,
$$

$$
\boldsymbol{U} = \overline{\boldsymbol{U}}, \text{ on } \Gamma_D,
$$

$$
\boldsymbol{P}\boldsymbol{n} = \boldsymbol{T}, \text{ on } \Gamma_N,
$$

where $\boldsymbol{f}$ is the body force, the displacement gradient $\nabla \boldsymbol{U}$ is a tensor field with the ij-component $\partial_{x_j} U_i$, $\operatorname{div} \boldsymbol{P}$ is a vector field with the i-component $\sum_j \partial_{x_j} P_{ij}$, $\mathbb{P}(\boldsymbol{K})$ is the nonlinear constitutive equation of the nonlinearly elastic body Ω, and $\boldsymbol{T}$ is the traction vector field with $T_i = \sum_j P_{ij} n_j$, on Γ_N.

The tensor fields $\boldsymbol{K}$ and $\boldsymbol{P}$ are not symmetric and are represented by 2×2 matrices which are not symmetric, in general. Let $[H(\operatorname{curl}; \Omega)]^2$ and $[H(\operatorname{div}; \Omega)]^2$ be the spaces of tensor fields such that each row of their matrix representation respectively belongs to $H(\text{curl}; \Omega)$ and $H(\text{div}; \Omega)$, where these latter spaces are defined in Example 2.9. Then, we can write the following mixed formulation for the above boundary value problem: Given a body force $\boldsymbol{f}$, a boundary displacement $\overline{\boldsymbol{U}}$, and a surface traction vector field $\boldsymbol{T}$ on Γ_N, find $\boldsymbol{U} \in [H^1(\Omega)]^2$, $\boldsymbol{K} \in [H(\text{curl}; \Omega)]^2$, and $\boldsymbol{P} \in [H(\text{div}; \Omega)]^2$ such that $\boldsymbol{U} = \overline{\boldsymbol{U}}$, on Γ_D, and

$$
\begin{aligned}
\int_\Omega \boldsymbol{P} : \nabla \boldsymbol{\Upsilon} &= \int_\Omega \boldsymbol{f} \cdot \boldsymbol{\Upsilon} + \int_{\Gamma_N} \boldsymbol{T} \cdot \boldsymbol{\Upsilon}, && \text{for all } \boldsymbol{\Upsilon} \in [H_D^1(\Omega)]^2, \\
\int_\Omega \nabla \boldsymbol{U} : \boldsymbol{\lambda} - \int_\Omega \boldsymbol{K} : \boldsymbol{\lambda} &= 0, && \text{for all } \boldsymbol{\lambda} \in [H(\text{curl}; \Omega)]^2, \qquad (5.28) \\
\int_\Omega \mathbb{P}(\boldsymbol{K}) : \boldsymbol{\pi} - \int_\Omega \boldsymbol{P} : \boldsymbol{\pi} &= 0, && \text{for all } \boldsymbol{\pi} \in [H(\text{div}; \Omega)]^2,
\end{aligned}
$$

where "$\cdot$" is the standard inner product and "$:$" is defined in (5.22).

Let the finite element spaces $[V_h]^2 \subset [H^1(\Omega)]^2$, $[V_h^c]^2 \subset [H(\text{curl}; \Omega)]^2$, $[V_h^d]^2 \subset [H(\text{div}; \Omega)]^2$ be the spaces respectively induced by 2-simplex of type (k), the Nédélec element, and the Raviart-Thomas element. Also let $[X_h]^2$ be the space of vector fields belonging to $[V_h]^2$ that vanish on Γ_D. Notice that $[V_h^c]^2$ and $[V_h^d]^2$ are simply two copies of the finite element spaces introduced in Sections 3.4.3 and 3.4.2. By using these finite element spaces and the weak form (5.28), we can obtain a conformal mixed finite element method for nonlinear elasticity. To solve this method, we have to solve a nonlinear system of equations since the constitutive equation $\mathbb{P}(\boldsymbol{K})$ is a nonlinear function of $\boldsymbol{K}$.

Newton's method provides an iterative framework to solve the above mixed method. The i-th Newton's iteration involves solving a linear problem and can be stated as follows: Given $\boldsymbol{U}_h^i \in [V_h]^2$ satisfying the boundary condition on Γ_D, $\boldsymbol{K}_h^i \in [V_h^c]^2$, and $\boldsymbol{P}_h^i \in [V_h^d]^2$, find $\boldsymbol{V}_h \in [X_h]^2$, $\boldsymbol{M}_h \in [V_h^c]^2$, and $\boldsymbol{Q}_h \in [V_h^d]^2$ and let

$$
\boldsymbol{U}_h^{i+1} = \boldsymbol{U}_h^i + \boldsymbol{V}_h, \quad \boldsymbol{K}_h^{i+1} = \boldsymbol{K}_h^i + \boldsymbol{M}_h, \quad \boldsymbol{P}_h^{i+1} = \boldsymbol{P}_h^i + \boldsymbol{Q}_h, \qquad (5.29)
$$

where $\boldsymbol{V}_h$, $\boldsymbol{M}_h$, and $\boldsymbol{Q}_h$ are obtained by solving the linear finite element method

$$
\begin{aligned}
\int_\Omega \boldsymbol{Q}_h : \nabla \boldsymbol{\Upsilon}_h &= \int_\Omega \{ -\boldsymbol{P}_h^i : \nabla \boldsymbol{\Upsilon}_h + \boldsymbol{f} \cdot \boldsymbol{\Upsilon}_h \} + \int_{\Gamma_N} \boldsymbol{T} \cdot \boldsymbol{\Upsilon}_h, && \text{for all } \boldsymbol{\Upsilon}_h \in [X_h]^2, \\
\int_\Omega \nabla \boldsymbol{V}_h : \boldsymbol{\lambda}_h - \int_\Omega \boldsymbol{M}_h : \boldsymbol{\lambda}_h &= \int_\Omega \{ -\nabla \boldsymbol{U}_h^i : \boldsymbol{\lambda}_h + \boldsymbol{K}_h^i : \boldsymbol{\lambda}_h \}, && \text{for all } \boldsymbol{\lambda}_h \in [V_h^c]^2, \\
\int_\Omega (\mathsf{A}(\boldsymbol{K}_h^i) : \boldsymbol{M}_h) : \boldsymbol{\pi}_h - \int_\Omega \boldsymbol{Q}_h : \boldsymbol{\pi}_h &= \int_\Omega \{ -\mathbb{P}(\boldsymbol{K}_h^i) : \boldsymbol{\pi}_h + \boldsymbol{P}_h^i : \boldsymbol{\pi}_h \}, && \text{for all } \boldsymbol{\pi}_h \in [V_h^d]^2,
\end{aligned}
$$

where $\mathsf{A}(\boldsymbol{K})$ is the elasticity tensor with the components A_{ijrs} and

$$
(\mathsf{A}(\boldsymbol{K}) : \boldsymbol{M}) : \boldsymbol{\pi} = \sum_{i,j,r,s} A_{ijrs} M_{rs} \pi_{ij}.
$$

Assume

$$n_1 = \dim[X_h]^2 = 2 \dim X_h,$$
$$n_c = \dim[V_h^c]^2 = 2 \dim V_h^c,$$
$$n_d = \dim[V_h^d]^2 = 2 \dim V_h^d,$$

and suppose $n_t = n_1 + n_c + n_d$. It is not hard to show that the above linear problem has a stiffness matrix with the structure

$$\mathbb{S}_{n_t \times n_t} = \begin{bmatrix} \mathbf{0} & \mathbf{0} & \mathbb{S}_{n_1 \times n_d}^{1d} \\ \mathbb{S}_{n_c \times n_1}^{c1} & \mathbb{S}_{n_c \times n_c}^{cc} & \mathbf{0} \\ \mathbf{0} & \mathbb{S}_{n_d \times n_c}^{dc} & \mathbb{S}_{n_d \times n_d}^{dd} \end{bmatrix}. \tag{5.30}$$

The above structure implies that the linear problem is not stable if either $n_d < n_1$ or $n_c < n_1$. As a consequence, the choices of 2-simplex of type (2) and the Nédélec element (of degree 1) or 2-simplex of type (2) and the Raviart-Thomas element (of degree 1) are not stable. The proof of these statements is left to the reader.

To demonstrate the implementation of this Newton's method for nonlinear elasticity in FEniCS, we consider compressible Neo-Hookean materials with the constitutive equation

$$\mathbb{P}(\boldsymbol{F}) = \mu \boldsymbol{F} - \mu \boldsymbol{F}^{-T} + 2\lambda(\ln I_3)\boldsymbol{F}^{-T}, \quad \mu, \lambda > 0,$$

with $\boldsymbol{F} = \boldsymbol{I} + \boldsymbol{K}$, where $\boldsymbol{I}$ is the identity matrix, and $I_3 = \det \boldsymbol{F}^T \boldsymbol{F}$. For this material, we have

$$A(\boldsymbol{K}) : \boldsymbol{M} = \mu \boldsymbol{M} + (\mu - 2\lambda \ln I_3)\boldsymbol{F}^{-T}\boldsymbol{M}^T\boldsymbol{F}^{-T} + 4\lambda(\operatorname{tr}\boldsymbol{F}^{-1}\boldsymbol{M})\boldsymbol{F}^{-T}.$$

We compute uniform compression of the 2D rectangular plate Ω shown in Figure 5.17, which is made of a Neo-Hookean material. A uniform vertical traction is applied on the top boundary. The horizontal displacement is assumed to be zero on the left and the right boundaries while the vertical displacement is assumed to be zero at the bottom boundary.

Deformations of the above 2D plate can be computed by using the following function.

```
def nonlinear_solver(n, degreeCG, degreeCurl, degreeDiv):

    # create mesh and define function space
    domain = Rectangle(dolfin.Point(0.0, 0.0),\
                       dolfin.Point(L, W))
    mesh = generate_mesh(domain,n)

    CGE   = FiniteElement("CG", mesh.ufl_cell(), degreeCG)
    CurlE = FiniteElement("N1curl", mesh.ufl_cell(), degreeCurl)
    DivE  = FiniteElement("RT", mesh.ufl_cell(), degreeDiv)
```

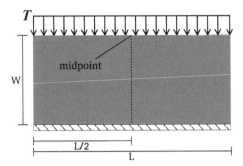

Figure 5.17 The geometry of the 2D plate example. The vertical displacement is zero at the bottom boundary and the horizontal displacement is zero on the left and the right boundaries.

```
Z = FunctionSpace(mesh,MixedElement([CGE, CGE, CurlE,\
                                     CurlE, DivE, DivE]))

# traction at the top boundary
Trac = Expression('alph * Lf', alph = 0.0, Lf = Lf, \
            degree = basedegree + degreeDiv)

boundary_parts = MeshFunction("size_t", mesh, \
                              mesh.topology().dim()-1)
# mark top facets as subdomain 1
class TopBoundary(SubDomain):
    def inside(self, x, on_boundary):
        tol = 1E-12  # tolerance for coordinate comparisons
        return on_boundary and abs(x[1] - W) < tol
Gamma_R = TopBoundary()
Gamma_R.mark(boundary_parts, 1)

# boundary functions
def left_func(x, on_boundary):
    tol = 1E-12     # tolerance for coordinate comparisons
    return on_boundary and abs(x[0]) < tol
def right_func(x, on_boundary):
    tol = 1E-12     # tolerance for coordinate comparisons
    return on_boundary and abs(x[0] - L) < tol
def bottom_func(x, on_boundary):
    tol = 1E-12     # tolerance for coordinate comparisons
    return on_boundary and abs(x[1]) < tol
bcs = [DirichletBC(Z.sub(0), Constant(0.0), left_func),
       DirichletBC(Z.sub(0), Constant(0.0), right_func),
       DirichletBC(Z.sub(1), Constant(0.0), bottom_func)]

# defining the constitutive equation and the elasticity tensor
def Pbb(K1, K2, Pi1, Pi2):
    # deformation gradient
```

```
        F = as_tensor([[1.0 + K1[0], K1[1]], [K2[0],\
                       1.0 + K2[1]]])
        PImat = as_tensor([[Pi1[0], Pi1[1]], \
                          [Pi2[0], Pi2[1]]])
        return inner(mu*F + (2*lam*ln(det(F.T*F)) \
                           - mu)*inv(F.T), PImat)

    def DPbb(K1, K2, M1, M2, Pi1, Pi2):
        # deformation gradient
        F = as_tensor([[1.0 + K1[0], K1[1]], [K2[0],\
                       1.0 + K2[1]]])
        FinvT = inv(F.T)
        Mmat = as_tensor([[M1[0], M1[1]], [M2[0], M2[1]]])
        PImat = as_tensor([[Pi1[0], Pi1[1]],\
                          [Pi2[0], Pi2[1]]])
        return inner(mu*Mmat + (mu - 2*lam*ln(det(F.T*F))) \
                    *FinvT*Mmat.T*FinvT+ 4*lam*\
                    tr(inv(F)*Mmat)*FinvT, PImat)

# obtaining initial guess, zero, by interpolating zero
initial_const = Constant((0.0, 0.0, 0.0, 0.0, 0.0, 0.0,\
                        0.0,0.0, 0.0, 0.0))
u_k = interpolate(initial_const, Z)
# _k variables only refer to the associated part of u_k
(U1_k, U2_k, K1_k, K2_k, P1_k, P2_k) = split(u_k)

# Newton iterations
(V1, V2, M1, M2, Q1, Q2) = TrialFunctions(Z)
(Up1, Up2, La1, La2, Pi1, Pi2) = TestFunctions(Z)

ds = Measure("ds", domain=mesh, \
             subdomain_data=boundary_parts)
LHS = ( inner(Q1, grad(Up1)) + inner(Q2, grad(Up2))
    + inner(grad(V1), La1) + inner(grad(V2), La2)
    - inner(M1, La1) - inner(M2, La2)
    + DPbb(K1_k, K2_k, M1, M2, Pi1, Pi2)
    - inner(Q1, Pi1) - inner(Q2, Pi2) )*dx
RHS = - ( inner(P1_k, grad(Up1)) + inner(P2_k, grad(Up2))
    + inner(grad(U1_k), La1) + inner(grad(U2_k), La2)
    - inner(K1_k, La1) - inner(K2_k, La2)
    + Pbb(K1_k, K2_k, Pi1, Pi2) - inner(P1_k, Pi1)
    - inner(P2_k, Pi2) )*dx  - Trac*Up2*ds(1)

z = Function(Z)         # increment vector
u = Function(Z)         # the current solution
omega = 0.8             # relaxation parameter
tol = 1.0E-5            # convergence tolerance
maxiter = 20            # maximum allowed iteration
```

```python
min_stepsize = 1.0E-4

alpha0 = 0.0              # adaptive loading parameter
u_alpha0 = Function(Z)    # the solution for alpha0
u_alpha0.assign(u_k)      # initiate u_alpha0
step_size = 1.0           # parameter for increasing alpha0

while alpha0 < 1.0:
    # alpha1 will be used to calculate the next step of loading
    alpha1 = alpha0 + step_size
    if alpha1 > 1.0:
        alpha1 = 1.0; step_size = 1.0 - alpha0

    print("-- load factor (alpha) = %6.2E -- " % alpha1)
    Trac.alph = alpha1     # increasing load gradually

    eps = 1.0             # infinity norm parameter
    iter = 0              # iteration counter
    while eps > tol and iter < maxiter:
        iter += 1
        if Guass_points:
            # number of points for Guassian Quadrature
            parameters["form_compiler"]["quadrature_degree"] =\
            Guass_points
        solve(LHS == RHS, z, bcs)

        eps = np.linalg.norm(z.vector().get_local(), ord=np.Inf)
        print("iter = %d: the infinity norm of the increment \
            is %8.2E" % (iter, eps))
        u.vector()[:] = u_k.vector() + omega*z.vector()
        u_k.assign(u)     # updating for the next iteration

    if iter == maxiter:
        print("Newton iterations reached the max iteration,\
            decreasing step-size (n = %d)" % n)
        step_size = 0.5 * step_size  # decreasing step_size
        u_k.assign(u_alpha0)         # rejecting the result
    else:
        u_alpha0.assign(u_k)         # accepting this step
        alpha0 = alpha1
        step_size = 2.0 * step_size  # increasing the step-size

    if step_size < min_stepsize:
        print("Too small step_size! step_size = %6.2E"\
            % step_size)
        import sys
        sys.exit("Termination!")
```

```
# extracting results
(U1, U2, K1, K2, P1, P2) = u_k.split()

# produce vtkfile for visualization of displacement
UFE = VectorElement(CGE)
W_u = FunctionSpace(mesh, UFE)
parameters['allow_extrapolation'] = True
U_vec = as_vector([U1, U2])
UU = project(U_vec, W_u)

return U2 , UU
```

To improve Newtons' iterations in the above function, instead of (5.29), the relations

$$U_h^{i+1} = U_h^i + \omega V_h, \quad K_h^{i+1} = K_h^i + \omega M_h, \quad P_h^{i+1} = P_h^i + \omega Q_h,$$

are used to find the approximate solution of the next step, where $0 < \omega \le 1$ is a relaxation parameter. Moreover, since the initial guess for the Newton method should be close to the solution, the load is gradually increased in `nonlinear_solver` to achieve large deformations.

The following program uses the function `nonlinear_solver` to compute deformations of the 2D plate. The displacement of the midpoint shown in Figure 5.17 is also calculated and outputs are saved in the VTK format, which is suitable for visualization in ParaView.

```
# degrees of CG, curl and div finite elements
degreeCG = 1; degreeCurl = 1; degreeDiv = 1

# number of divisions
n = 7

# length and width
L = 15.0 ; W = 10.0

# material properties
mu = 0.0002; lam = 1.0

# basedegree will be added to degree of the associated FE
basedegree = 2

# input loads
loadings = [0.25, 1.25, 2.5, 4.5, 9.5, 25, 125, 250]

# midpoint displacement vector
u_mid = np.zeros((len(loadings),))
```

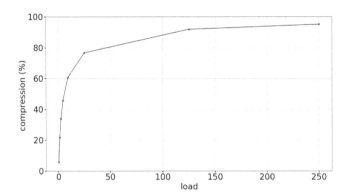

Figure 5.18 Compression of the 2D plate versus load.

```
# number of Gaussian points
Guass_points = 10

for (i, Lf) in enumerate(loadings):

    # solver
    U2, UU = nonlinear_solver(n, degreeCG, \
                              degreeCurl, degreeDiv)

    # record displacement of midpoint
    u_mid[i] = U2(L/2, W)

    # print compression after each iteration
    print("compression (%) = ", (u_mid/-W)*100)

# save vtkfile
vtkfile = File('Ch5_NonlinearElas/disp.pvd')
vtkfile << UU
```

Figure 5.18 shows the compression versus load diagram for the 2D plate. This figure is obtained by using the parameters given in the above programs and the compression is defined as the vertical displacement of the midpoint divided by the height of the the plate. Figure 5.19 depicts a reference mesh of the 2D plate together with some deformed configurations. Colors in the deformed configurations indicate the distribution of the displacement norm.

Figure 5.19 A reference mesh together with some deformed configurations of the 2D plate with 5.7%, 33.8%, and 60.6% compression.

EXERCISES

Exercise 5.1. Derive the weak formulation (5.2) of elastic bars by using the strong form (5.1).

Exercise 5.2. Derive a weak formulation for the elastic bar of Section 5.1 assuming $f = 0$, and fixed end points, that is, $u(0) = u(l) = 0$.

Exercise 5.3. For the bar of Exercise 5.2 obtain the exact solution u if E, A, and b are constant. Show that the finite element solution corresponding to a mesh with only one element and 1-simplex of type (2) is equal to the exact solution. Is this result holds if one uses 1-simplex of type (1)? Justify your answer.

Exercise 5.4. Derive a weak formulation for the elastic bar of Section 5.1 subject to the axial distributed load $b(x)$ and the axial end loads f_0 and f_l at $x = 0$, and $x = l$, respectively. Is the solution of this problem unique? Also can we choose the values of f_0, f_l, and $b(x)$ independently?

Exercise 5.5. Use the Euler-Bernoulli theory, the Hermit 1-simplex of type (3), and a mesh consisting of only 1 element to derive the finite element approximation of the deflection $u(x)$ of the cantilever beam of Figure 5.2 with $f = 0$ and $Q = 0$. What is the associated L^2-error?

Exercise 5.6. Repeat Exercise 5.5 assuming that $Q = 0$, $M = 0$, and f is a constant function.

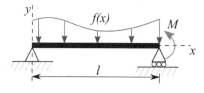

Figure 5.20 A simply-supported beam under a distributed load $f(x)$ and a moment M.

Exercise 5.7. Repeat Exercise 5.5 for the simply-supported beam shown in Figure 5.20 assuming that $f = 0$. Notice that $u(0) = u(l) = 0$, and the bending moment is 0 and M at $x = 0$ and $x = l$, respectively.

Exercise 5.8. Repeat Exercise 5.7 assuming that $M = 0$, and f is a constant function.

Exercise 5.9. Derive the mixed formulation (5.11) for cantilever beams.

Exercise 5.10. Derive a mixed formulation similar to (5.11) for the simply-supported beam of Figure 5.20.

Exercise 5.11. Derive the weak formulation (5.14) and the finite element method (5.15) associated to the wave equation.

Exercise 5.12. Derive the mixed formulation (5.21) of the Stokes equation.

Exercise 5.13. Another boundary condition for the Stokes equation can be described as follows: Suppose the boundary of the domain Ω can be written as $\partial\Omega = \Gamma_D \cup \Gamma_N$, and consider the boundary condition

$$\begin{cases} \boldsymbol{v} = \boldsymbol{v}_0, & \text{on } \Gamma_D, \\ \mu\nabla\boldsymbol{v}\cdot\boldsymbol{n} - p\boldsymbol{n} = \boldsymbol{g}, & \text{on } \Gamma_N, \end{cases}$$

where $\boldsymbol{n}$ is the unit outward normal at $\partial\Omega$, the i-th component of the vector field $\nabla\boldsymbol{v}\cdot\boldsymbol{n}$ is given by $\sum_{j=1}^{n} n_j \partial_{x_j} v_i$, and $\boldsymbol{v}_0$ and $\boldsymbol{g}$ are given vector fields. Derive the following weak form of the Stokes equation subject to the above boundary condition: Find $\boldsymbol{v} \in [H^1(\Omega)]^n$ and $p \in P$ such that $\boldsymbol{v} = \boldsymbol{v}_0$, on Γ_D, and

$$\int_\Omega \mu\nabla\boldsymbol{v} : \nabla\boldsymbol{w} - \int_\Omega p\,\text{div}\,\boldsymbol{w} = \int_\Omega \boldsymbol{f}\cdot\boldsymbol{w} + \int_{\Gamma_N} \boldsymbol{g}\cdot\boldsymbol{w}, \quad \text{for all } \boldsymbol{w} \in [H_D^1(\Omega)]^n,$$

$$-\int_\Omega q\,\text{div}\,\boldsymbol{v} = 0, \qquad\qquad\qquad \text{for all } q \in P, \tag{5.31}$$

where $[H_D^1(\Omega)]^n$ is the space of vector fields that vanish on Γ_D. Unlike the problem (5.21), the problem (5.31) has a unique solution and there is no need to manipulate pressure.

Exercise 5.14. Derive the weak formulation (5.26) for linearized elasticity.

Exercise 5.15. Derive the implicit finite element method (5.27) for linearized elastodynamics.

Exercise 5.16. Show that the stiffness matrix (5.30) implies that the choice of 2-simplex of type (2) and the Nédélec element leads to unstable Newton's iterations for nonlinear elasticity.

Exercise 5.17. Show that the stiffness matrix (5.30) implies that the choice of 2-simplex of type (2) and the Raviart-Thomas element leads to unstable Newton's iterations for nonlinear elasticity.

COMPUTER EXERCISES

Computer Exercise 5.1. Consider the elastic bar of Section 5.1 with $EA = 1$, $l = 2$, $f = e^1$, $b(x) = -\frac{EA}{l^2}e^{\frac{x}{l}}$, and $\partial_x u|_{x=l} = \frac{fl}{EAe}$. Show that the exact solution of this problem is $u(x) = e^{\frac{x}{l}} - 1$, and compute L^2-errors and convergence rates associated to a family of meshes and 1-simplices of type (k), with $k = 1, 2$.

Computer Exercise 5.2. Modify the program of Section 5.2 to approximate the solution of Exercise 5.5 by using a mixed finite element method. Use the mesh sizes $N = 3, 6, 9$, and the data $EI = 1$, $l = 1$, and $M = 1$. Compute the associated L^2-errors and the convergence rate of the deflection.

Computer Exercise 5.3. Repeat Computer Exercise 5.2 for Exercise 5.6 with $f = 1$.

Computer Exercise 5.4. Repeat Computer Exercise 5.2 for Exercise 5.7 with $M = 1$.

Computer Exercise 5.5. Repeat Computer Exercise 5.2 for Exercise 5.8 with $f = 1$.

Computer Exercise 5.6. Derive an explicit finite element method for the initial-boundary value problem (5.14) and implement your method in FEniCS. By developing a test problem, compare your results with those based on the implicit method (5.15).

Computer Exercise 5.7. Modify the code of Section 5.4 to model free vibrations of a membrane in the following sense: Using the notation of that section, assume the membrane $\Omega = (0, L) \times (0, W)$ is fixed on its boundary and consider the initial condition $\partial_t u = 0$, and $u(x, y, 0) = \sin \frac{k\pi x}{L} \sin \frac{m\pi y}{W}$, where k and m are arbitrary integers.

Computer Exercise 5.8. Is 2-simplex of type (2) and the Raviart-Thomas element a stable choice for the mixed formulation (5.18)? Justify your answer by using the structure of the associated stiffness matrix.

Computer Exercise 5.9. Write a function similar to `seepage_solver` of Section 5.6 that implements the mixed formulation (5.18) with $V = [H^1(\Omega)]^2$. Try your code with degree 2 elements for head and degree 1 elements for discharge and compare your result with that of Computer Exercise 5.8.

Computer Exercise 5.10. Implement the weak form of Exercise 5.13 for the Stokes equation by modifying the code of Section 5.8. Debug your code by designing a test problem and computing the convergence rates of velocity and pressure.

Computer Exercise 5.11. Modify the code of Section 5.9 to compute the deformation of the underlying beam under its own weight. Assume the beam is made of steel and has a uniform density.

Computer Exercise 5.12. Modify the code of Section 5.9 to compute the deformation of the underlying beam under a uniform distributed load in the y-direction at its top boundary. Assume the net load due to this distributed load is equal to the weight of the beam given in Computer Exercise 5.11.

Computer Exercise 5.13. Modify the program of Section 5.10 to compute stress of the specimen. Plot the L^2-norm of stress versus time, where the square of the L^2-norm of stress is equal to the summation of the squares of the L^2-norm of stress components.

Computer Exercise 5.14. Run the program of Section 5.11 by using 2-simplex of type (2). Does your solution converge? Justify your answer by using Exercises 5.16 and 5.17.

Computer Exercise 5.15. Extend the program of Section 5.11 to the 3D case.

A Installation of FEniCS

FEniCS can be installed on different operating systems. All programs of this book were tested with FEniCS version 2019.1.0 on Linux Ubuntu. Here, we mention how to install FEniCS on Ubuntu. The latest information on available options for FEniCS installation including Windows and Mac installations can be found at the official FEniCS web page.[1] Windows and Mac users can also employ the following instructions on a virtual machine[2] running Ubuntu.

There are two approaches for installing FEniCS on Ubuntu: Ubuntu personal package archives (PPA) and Ubuntu repositories. Unlike the latter, the former always contains the latest release of FEniCS. For the PPA installation, copy the following lines in an Ubuntu shell window:[3]

```
$ sudo apt-get install --no-install-recommends \
    software-properties-common
$ sudo add-apt-repository ppa:fenics-packages/fenics
$ sudo apt-get update
$ sudo apt-get install --no-install-recommends fenics
```

For the installation using repositories, use the following lines:

```
$ sudo apt-get update
$ sudo apt-get install --no-install-recommends fenics
```

To verify the installation, enter the following code in a FEniCS-enabled window, which should run without any error and will produce no output.

```
$  python -c 'import fenics'
```

---

[1] https://fenicsproject.org

[2] https://www.virtualbox.org

[3] In this and the next appendix, $ at the beginning of a line indicates its execution in an Ubuntu shell window.

B Introduction to Python

A brief introduction to Python is presented in this appendix to help the readers who are not familiar with Python. We assume that the readers are familiar with basic programming skills. The goal is to provide some basic knowledge sufficient to run and understand the main aspects of the programs of this book. This introduction is by no means complete and for further information, the readers are referred to abundant resources on Python programming such as the official Python web page[1] and References [8, 12].

The latest version of Python and complete instructions for Python installations on different operating systems are available at the official Python web page.[2] Programs of this book are based on Python 3 and were tested on Linux Ubuntu. Python is installed on most Linux distributions by default. On an Ubuntu machine, type the following in a shell window to verify the Python version.

```
$ python3 --version
```

To install the latest version of Python[3] on an Ubuntu machine, simply type the following commands:

```
$ sudo apt-get update
$ sudo apt-get install python3.8
```

B.1 RUNNING PYTHON PROGRAMS

Python is an interpreted language in the sense that Python scripts are executed without being compiled beforehand. Two common approaches for developing and running Python programs are:

1. Using add-on packages such as the Jupyter Notebook[4] and Spyder[5]. The Jupyter Notebook provides an interactive environment for developing Python codes. Jupyter notebooks containing all programs of this book can be downloaded from the companion web page of this book.
2. Simply using a text editor and a terminal window.

Executing Python codes on an Ubuntu machine is straightforward. To obtain a simple interactive mode type:

---

[1]https://docs.python.org/3/tutorial/index.html

[2]https://www.python.org/downloads/

[3] At the time of this publication, the latest version of Python is 3.8.

[4]https://jupyter.org/

[5]https://www.spyder-ide.org/

```
$ python3
```

This will open a Python shell similar to the following:

```
Python 3.8.0 (default, Oct 28 2019, 16:14:01)
[GCC 8.3.0] on linux
Type "help", "copyright", "credits" or "license" for more
information.
>>>
```

Python commands are entered after >>>. In the followings, >>> at the beginning of a line indicates its execution in an interactive environment such a Python shell or a Jupyter notebook. To exit a Python shell type exit().

A Python shell is suitable for examining simple codes. Assume that the file Test.py contains a Python program possibly containing several thousands of lines. To run this program, one types:

```
$ python3 Test.py
```

Files containing Python programs can be prepared by plain text editors or more advanced editors such as Spyder and PyCharm[6].

B.2   LISTS

A simple list can be defined as:

```
>>> N = [4, 8, 12, 16, 20]
```

The expression N[i-1] specifies the i-th element of a list. For example, the first and the last elements of N can be printed as follows:

```
>>> print('first element =', N[0])
first element = 4
>>> print('last element =', N[4])
last element = 20
```

Notice that the index of Python lists starts at 0. An alternative way to achieve the above results is

```
>>> print('first element =', N[-5])
first element = 4
>>> print('last element =', N[-1])
last element = 20
```

---

[6]https://www.jetbrains.com/pycharm/

It is also possible to access multiple elements of lists. For example, the following code prints the first three elements of N.

```
>>> print('first three elements are', N[0:3])
first three elements are [4, 8, 12]
```

One can add an element to the end of a list:

```
>>> N.append(24)
>>> print(N)
[4, 8, 12, 16, 20, 24]
```

The number of elements of a list are determined by len():

```
>>> print(len(N))
6
```

It is possible to assign new values to elements of a list as follows:

```
>>> N[0] = -100
>>> print(N)
[-100, 8, 12, 16, 20, 24]
>>> N[0:3] = [-1, -2, -3]
>>> print(N)
[-1, -2, -3, 16, 20, 24]
```

B.3   BRANCHING AND LOOPS

Python has the standard if-else and if-elif-else blocks to branch the flow of programs based on given conditions. Here the keyword elif means "else if". A simple example is given below.

```
>>> A = 21; B = 11
>>> if A > B:
...        print("A is greater than B")
... elif A < B:
...        print("A is smaller than B")
... else:
...        print("A is equal to B")

A is greater than B
```

Notice that unlike many programming languages, there is no keywords to specify the end of the block. In Python, lines with the same indentation level are considered to belong to the same block. Therefore, it is very important to pay attention to the indentation of each line when developing codes.

The standard for and while loops are also available. For example, the following lines print elements of a list by using these loops.

```
>>> M = [1, 0.5, -5]
>>> for i in range(len(M)):
...        print(M[i])
...
1
0.5
-5
>>> i = 0
>>> while i < len(M):
...        print(M[i])
...        i += 1    # update i
...
1
0.5
-5
```

Here i += 1 is equivalent to i = i + 1. Also notice that comments begin with # in Python.

B.4   FUNCTIONS

A function is a tool for grouping statements such that they can be easily called several times in a program. In Python, the definition of a function begins with def and results are sent back using return. As an example, consider the step function

$$f(x) = \begin{cases} 1, & \text{if } x \geq 0, \\ 0, & \text{otherwise.} \end{cases}$$

The following lines define this function and compute $f(0.5)$.

```
>>> def f(x):
...        if x >= 0:
...                y = 1
...        else:
...                y = 0
...        return y
...
>>> f(0.5)
1
```

As a general rule, all variables defined in a function are local to that function and they will not exist anymore after the function returns its value.

B.5   CLASSES AND OBJECTS

Python is an object-oriented programming language and extensively employs classes. Here we give a simple example of a class in Python. Suppose that we want to define the function

$$g(x,y) = \left(x^2 + y^2\right)^b + t^c,$$

where a, b, and t represent some physical parameters such as time that may change. To define this function, one can use the following class G that consists of the variables b,c,t, and also the function value that returns the value of g at (x,y):

```
>>> class G(object):
...       def __init__(self, b, c, t):
...              self.b = b
...              self.c = c
...              self.t = t
...       def value(self, x, y):
...              v = pow(x**2 + y**2, self.b) \
...                     + self.t**self.c
...              return v
...
```

Suppose $b = c = 1$, and $t = 0$. Now, the function g can be defined as an object, also called an instance, of the class G:

```
>>> g = G(1,1,0)
```

To calculate $g(2,3)$, we use the statement

```
>>> g.value(2,3)
13
```

Assume that the value of t changes to 1 and we want to compute $g(2,3)$ with this new value of t. This can be achieved by using the following lines:

```
>>> g.t = 1
>>> g.value(2,3)
14
```

Functions and variables of a class are commonly called methods and data attributes, respectively. In the above example, G has the method value and the data attributes b,c,t.

FEniCS has many classes. To be able to use them, one needs to import them at the beginning of a program:

```
>>> from dolfin import *
```

This way, all classes of FEniCS become available. For example, we can use the class UnitSquareMesh to define a 10×10 mesh of a unit square:

```
>>> mesh = UnitSquareMesh(10, 10)
```

The class `UnitSquareMesh` has several data attributes and methods. To see them, simply type `mesh.` and press the tab bottom. This will show a long list that begins with

```
>>> mesh.
mesh.bounding_box_tree(          mesh.num_entities(
mesh.cell_name(                  mesh.num_entities_global(
mesh.cell_orientations(          mesh.num_faces(
mesh.cells(                      mesh.num_facets(
mesh.color(                      mesh.num_vertices(
mesh.coordinates(                mesh.order(
```

As an example, we can employ the method hmax to calculate the maximum diameter of elements h:

```
>>> mesh.hmax()
0.14142135623730964
```

B.6   READING AND WRITING FILES

Assume that we have the text file `Data.txt` that contains velocities of some points of a mechanical system as follows:

```
Point    Velocity(m/s)
1        5.5
2        6.7
3        6.3
4        9.1
5        1.0
6        3.0
```

Suppose we want to calculate the average of velocities and save it in another text file `Result.txt`. First, we read `Data.txt` line by line and calculate the average average:

```
>>> data = open('Data.txt', 'r') # open the file for reading
>>> data.readline()      # skip the first line
'Point\tVelocity(m/s)\n'
>>> Num = 0; sum = 0
>>> for line in data:
...      line_list = line.split() # list of strings for each line
...      velocity = float(line_list[1])
...      sum += velocity; Num += 1
...
>>> data.close() # close the file
>>> average = sum / Num
```

```
>>> average
5.266666666666667
```

The line data = open('Data.txt', 'r') opens Data.txt for reading. Since the first line does not contain any velocity, we skip it by using data.readline(). In the for loop, we read the file line by line. The code line_list = line.split() splits the string line into words separated by blanks and creates the list of strings line_list of these words. For example, in the first iteration, we have line_list = ['1', '5.5']. To extract the velocity as a number, we select the second element of line_list and convert it to a floating point number by the line velocity = float(line_list[1]). After the for loop, we first close the data file and then we compute the average.

To write the result, we create Result.txt and open it for writing:

```
>>> result = open('Result.txt', 'w')
```

Then, we write the result and close the file.

```
>>> result.write('average velocity = %.8f' % average)
>>> result.close()
```

This will create the text file Result.txt that contains

```
average velocity = 5.26666667
```

B.7 NUMERICAL PYTHON ARRAYS

Numerical Python, abbreviated as NumPy, is a Python package designed for scientific computations. In NumPy, we employ arrays instead of lists to store vectors and matrices. To define a NumPy array, we first need to import NumPy. The common statement for this purpose is

```
>>> import numpy as np
```

Numpy arrays can be defined similar to lists. For example, to define the matrix

$$M = \begin{bmatrix} 4 & 8 & 12 \\ 16 & 20 & 24 \end{bmatrix},$$

we employ the line

```
>>> M = np.array([[4, 8, 12], [16, 20, 24]])
>>> M
array([[ 4,  8, 12],
       [16, 20, 24]])
```

Elements of an array are accessed similar to lists:

```
>>> M[0] [0]
4
>>> M[0,0]
4
```

NumPy has several tools to define more complex arrays. For example, an array for a uniform division of the interval $[-2,2]$ to 4 sub-intervals is obtained by

```
>>> np.linspace(-2,2,5)
array([-2., -1.,  0.,  1.,  2.])
```

A 2×2 zero matrix and a 3×2 matrix of ones can be defined as:

```
>>> np.zeros((2,2))
array([[0., 0.],
       [0., 0.]])
>>> B = np.ones((3,2))
>>> B
array([[1., 1.],
       [1., 1.],
       [1., 1.]])
```

Different operations for vectors and matrices are also available. For example, the matrix product $M \times B$ of the above matrices is computed as

```
>>> np.dot(M,B)
array([[24., 24.],
       [60., 60.]])
```

An important feature of Numpy is *vectorization*, that is, instead of using loops over arrays, many NumPy operations can be directly preformed on arrays. Vectorization is useful for speeding up heavy numerical computations. As an example, suppose that we want to compute $f(x) = e^x \sin(x^2)$, where x belongs to a uniform mesh of $[0,1]$ with $2,000,000$ vertices. In the following, we preform a standard loop version and a vectorized version of this computation and compare the time spent on each version.

```
>>> import time
>>> import math
>>> import numpy as np
>>> x = np.linspace(0,1,2000000)
>>> Time_Start = time.time();\
... f_for = [math.exp(x[i])*math.sin(x[i]**2) \
...            for i in range(len(x))]; \
... print("***** for version: %.8f seconds"
...            % (time.time() - Time_Start))
***** for version: 2.00471950 seconds
```

```
>>> Time_Start = time.time();\
... f_vec = np.exp(x)*np.sin(x**2);\
... print("***** vectorized version: %.8f seconds"
...              % (time.time() - Time_Start))
***** vectorized version: 0.14114642 seconds
```

B.8 PLOTTING WITH MATPLOTLIB

There are several Python packages for plotting and visualization such as Matplotlib, Mayavi, and Easyviz. Matplotlib is the standard package for plotting curves. The common statement for importing this package is

```
>>> import matplotlib.pyplot as plt
```

We plot two simple examples in this section. First, we plot the function $\sin(x)$ on $[0, 2\pi]$. This can be achieved by the following lines:

```
>>> import numpy as np
>>> x = np.linspace(0, 2*np.pi, 40)
>>> y = np.sin(x)
>>> plt.plot(x,y)
[<matplotlib.lines.Line2D object at 0x7fb6300e6e10>]
>>> plt.show()
```

These lines produce a figure similar to Figure B.1.

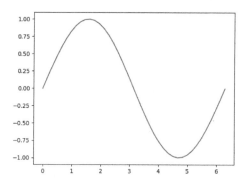

Figure B.1 Plot of $\sin(x)$ on $[0, 2\pi]$.

Next, we simultaneously plot the functions $\sin(x)$ and $\cos(x)$ on $[0, 2\pi]$ and add some information to the figure:

```
>>> x = np.linspace(0, 2*np.pi, 40)
```

```
>>> S = np.sin(x)
>>> C = np.cos(x)
>>> plt.plot(x, S, '-bo', linewidth = '4', markersize = '8')
[<matplotlib.lines.Line2D object at 0x7ff5e18505c0>]
>>> plt.plot(x, C, '-rs', linewidth = '4', markersize = '8')
[<matplotlib.lines.Line2D object at 0x7ff5e18507b8>]
>>> plt.xlabel('x', fontsize = '14')
Text(0.5, 0, 'x')
>>> plt.ylabel('y', fontsize = '14')
Text(0, 0.5, 'y')
>>> plt.legend(["sin(x)","cos(x)"], loc = 'best')
<matplotlib.legend.Legend object at 0x7ff5e1850e48>
>>> plt.savefig('Figure_2.png')
>>> plt.show()
```

These lines are self-explanatory and yield a figure similar to Figure B.2. Moreover, the line plt.savefig('Figure_2.png') saves the figure as Figure_2.png.

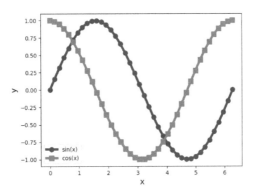

Figure B.2 Plots of $\sin(x)$ and $\cos(x)$ on $[0, 2\pi]$.

References

1. R.A. Adams and J.J.F. Fournier. *Sobolev Spaces*. Academic Press, Amsterdam, 2003.
2. A. Angoshtari and A. Gerami Matin. A conformal three-field formulation for nonlinear elasticity: From differential complexes to mixed finite element methods. *arXiv preprint arXiv:1910.09025*, 2020.
3. R.G. Bartle and D.R. Sherbert. *Introduction to Real Analysis*. John Wiley & Sons, Inc. New York, USA, 2011.
4. D. Boffi, F. Brezzi, and M. Fortin. *Mixed Finite Element Methods and Applications*. Springer-Verlag, Berlin, 2013.
5. P.G. Ciarlet. *The Finite Element Method for Elliptic Problems*, volume 40. SIAM, Philadelphia, USA, 2002.
6. A. Ern and J. Guermond. *Theory and Practice of Finite Elements*. Springer-Verlag, New York, 2004.
7. L.C. Evans. *Partial Differential Equations*. American Mathematical Society, Providence, RI, 2010.
8. H.P. Langtangen. *A Primer on Scientific Programming with Python*, volume 6. Springer, Berlin, 2016.
9. H.P. Langtangen and A. Logg. *Solving PDEs in Python: The FEniCS Tutorial*. Springer, 2017.
10. A. Logg, K.A. Mardal, and G. Wells. *Automated Solution of Differential Equations by the Finite Element Method: The FEniCS Book*, volume 84. Springer Science & Business Media, 2012.
11. A.W. Naylor and G.R. Sell. *Linear Operator Theory in Engineering and Science*. Springer Science & Business Media, 2000.
12. B. Slatkin. *Effective Python: 90 Specific Ways to Write Better Python*. Addison-Wesley Professional, 2019.

Index

abstract problem, 60
    coercivity, 61, 64
    ellipticity, 61
    variational structure, 61
    well-posedness, 60, 64
advection-diffusion equation, 56, 108
affine family of finite elements, 46
affine mapping, 8
affine-equivalent, 45
approximability property, 65
approximation space, 19

Babuška-Brezzi condition, 124
barycenter, 27
barycentric coordinates, 27
basis (Hamel), 9
bilinear form, 9
boundary condition, 56
    Dirichlet, 57
    essential, 58
    mixed Dirichlet-Neumann, 59
    natural, 59
    Neumann, 58
    Robin, 59
boundary value problem, 56
bounded set, 5
    from above, 5
    from below, 5
Box, 36

Céa's lemma, 65
CG, 30
conformal space, 38
conforming method, 62
consolidation problem, 117
conventional finite element diagram, 28
convergence, 46, 64
convergence rate, 23, 47, 65
    optimal, 24, 47, 66
curl, 8

degree of freedom
    global, 19, 38
    local, 25, 26
diameter of element, 35
diffusion problem, 79
dimension, 9
DirichletBC, 68, 73, 87
divergence div, 7
dolfin, 22
ds, 68, 73
dual basis, 14
dual space, 13
duality brackets $\langle \cdot, \cdot \rangle$, 14
dx, 68

elastic membrane, 56, 104
elasticity
    linearized, 56, 126, 130
    nonlinear, 134
elliptic PDE, 55
error, 46, 64
errornorm, 23
Euler method, 79
    explicit, 79
    implicit, 79
existence of solution, 56
Expression, 22

FEniCS, 21, 66
    installation, 147
fenics, 22
finite element, 26
    assembly process, 25, 37
    Crouzeix-Raviart, 50
    Hermit, 30
    Hermite n-simplex of type (3), 30
    Lagrange, 25, 27
    Nédélec, 33
    node, 27, 30
    Raviart-Thomas, 32

simplex of type (k), 29
  space, 37
finite element method, 62
  conforming, 62
  mixed, 84, 123
finite-dimensional, 9
FiniteElement, 22, 41
full rank, 10, 85
function, 5
  continuous $C^m(\Omega), C^m(\overline{\Omega})$, 6
  bijective, 6
  domain, 5
  extension, 5
  invertible, 6
  one-to-one, injective, 6
  onto, surjective, 6
  range, 6
  restriction, 5
functional, 13
FunctionSpace, 22, 41, 87

Gaussian elimination, 69
Gelerkin method, 62
generate_mesh, 36
grad, 68
gradient ∇, 14
Green's formulas, 14

h-type approach, 50
hat function, 20
heat transfer equation, 56, 77, 82
hyperbolic PDE, 92

inf-sup condition, 85
infimum, 5
infinite-dimensional, 9
initial condition, 77, 92
initial-boundary value problem, 77, 79,
    92, 105
inner, 68
inner product, 14
interpolant, 26
  global, 38
  Lagrange, 19, 27, 41
  local, 24

Nédélec, 45
  Raviart-Thomas, 44
interpolate, 23, 41
interpolation operator, 21
  global I_h, 38
  local I_K, 26
  Nédélec I_K^N, I_h^N, 34, 45
  Raviart-Thomas I_K^{RT}, I_h^{RT}, 32, 44
  simplex of type (k) I_K^k, I_h^k, 29, 40
isoparametric family of finite elements,
    46
iterative method, 69

kernel, 8
Kronecker delta δ_{ij}, 14

Lagrange, 22, 25, 30
Laplacian Δ, 15, 122
Lax-Milgram lemma, 61
Lebesgue space $L^2(\Omega)$, 7
linear combination, 9
linear mapping, 8
linear space, 6
linear subspace, 8
linearly independent, 9
locally supported, 38
lower bound, 5

mass matrix, 78
maximum, 5
Measure, 73
mesh, 19, 34
  affine, 35
  cell, 34
  element, 19, 34
  geometrically conformal, 35
  number of edges n_e, 35
  number of elements n_{el}, 35
  number of faces n_f, 35
  number of vertices n_v, 35
  number of vertices on boundary n_v^∂,
    85
  vertex, 19
MeshFunction, 72
method of lines, 78

minimum, 5
mixed formulation, 83, 102, 123
MixedElement, 87
mshr, 36

N1curl, 34
nodal basis, 27
non-conforming method, 62
norm, 12
H^1, $\|\cdot\|_{1,2}$, 12
L^2, $\|\cdot\|_2$, 12
$\|\cdot\|_c$, 13
$\|\cdot\|_d$, 13
normal derivative ∂_n, 15
normed linear space, 12
null space, 8
Numerical Python, 155
NumPy, 155

on_boundary, 68

p-type approach, 50
parabolic PDE, 77
partly Sobolev class, 7
$H(\mathrm{curl};\Omega)$, 8
$[H(\mathrm{curl};\Omega)]^n$, 135
$H(\mathrm{div};\Omega)$, 7
$[H(\mathrm{div};\Omega)]^n$, 135
Point, 36
Poisson's equation, 56, 66, 85, 104
Polygon, 36
polynomial space
$\mathbb{NE}$, 33
$\mathbb{P}_k(\mathbb{R}^n)$, 7
$\mathbb{Q}_k(\mathbb{R}^n)$, 16
$\mathbb{RT}$, 31
positive definite matrix, 64
preconditioned Krylov solver, 69
project, 90
projection, 83, 89
Python programming language, 149

rank-nullity theorem, 10
real numbers, 5
Rectangle, 36

reference element, 35
reference finite element, 45
regularity of solution, 56
Ritz method, 62
RT, 33

saddle-point variational structure, 84, 123
seepage problem, 112
set
closure, 6
open, 6
shape function
global, 21, 38
local, 25, 26
shape-regular, 46
simplex, 27
center of gravity, 27
edge, 27
face, 27
vertex, 27
size_t, 72
Sobolev space, 7
$H^m(\Omega)$, 7
$[H^1(\Omega)]^n$, 7
$H_0^1(\Omega)$, 8
$[H_0^1(\Omega)]^n$, 122
$H_D^1(\Omega)$, 59
$[H_D^1(\Omega)]^n$, 127
solution space, 57, 62
solve, 69
span of a set, 9
sparse LU decomposition, 69
sparse matrix, 63
Sphere, 36
split(), 88
stiffness matrix, 63
time-dependent, 78
Stokes equation for fluids, 144
strong solution, 57
SubDomain, 72
supremum, 5

Taylor-Hood element, 124
test function, 57, 62

test problem, 66
test space, 57, 62
TestFunction, 68
TestFunctions, 87
the global degrees of freedom, 38
three-point centered-difference formula,
        92
    implicit, 106, 130
time step, 79
trial space, 57, 62
TrialFunction, 68
TrialFunctions, 87
triangulation, 35

UFL, 68
uniqueness of solution, 56
unit n-simplex, 27, 51
unit ball, 6
unit cube, 36
unit sphere, 6
unit square, 15, 36
UnitCubeMesh, 36
UnitIntervalMesh, 22
UnitSquareMesh, 36
upper bound, 5

variational crimes, 94
variational structure, 61
vector space, 6
VectorFunctionSpace, 42
vectorization, 156

wave equation, 93, 105
weak formulation, 56, 57
weak solution, 56, 57
well-posedness, 60, 64